Mammals of the Carolinas, Virginia, and Maryland

The Fred W. Morrison Series
in Southern Studies

Mammals of the Carolinas, Virginia, and Maryland

Wm. David Webster,
James F. Parnell, and
Walter C. Biggs, Jr.

Photographs by
James F. Parnell

The University of
North Carolina Press
Chapel Hill and London

Library of Congress Cataloging in Publication Data

Webster, William David.
    Mammals of the Carolinas, Virginia, and Maryland.

    Bibliography: p.
    Includes index.
        1. Mammals—Middle Atlantic States.   I. Parnell,
James F.   II. Biggs, Walter C., Jr.   III. Title.
    QL719.M54W43   1985   599.0975   85-1171
    ISBN 0-8078-1663-9
    ISBN 0-8078-5542-1 (pbk. : alk paper)

cloth 08 07 06 05 04   5 4 3 2 1

paper 08 07 06 05 04   5 4 3 2 1

Contents

Acknowledgments

This book could not have been produced without the assistance of many colleagues and friends. We gratefully acknowledge research performed over the years by many mammalogists who have written extensively about details of the lives of mammals occurring in the Carolinas, Virginia, and Maryland. We relied heavily upon their work; many of their publications are listed in the references section of this book.

Several persons read portions of the manuscript and in some cases provided us with results of their ongoing, but as yet unpublished, research. These include Dr. Joseph A. Chapman, Utah State University; Mr. William Cook, Great Smoky Mountains National Park; Mr. David S. Lee, North Carolina State Museum of Natural History; Dr. John F. Pagels, Virginia Commonwealth University; Dr. John L. Paradiso, U.S. Fish and Wildlife Service; Dr. Roger A. Powell, North Carolina State University; Dr. Robert K. Rose, Old Dominion University; Dr. William R. Teska, Furman University; and Dr. Peter D. Weigl, Wake Forest University. Their comments and suggestions have been invaluable; however, errors of omission and commission inevitably remain, for which we alone are responsible.

We are grateful to the following persons who assisted us in a variety of ways in securing mammal specimens to photograph: Ms. Mary K. Clark, Mr. Peter B. Colwell, Mr. Robert R. Currie, Mr. David S. Lee, Mr. John J. Lepri, Dr. David G. Lindquist, Mr. Edwin H. Manchester, Mr. Claude H. McAllister, Mr. Jim Midkiff, Mr. Mark A. Shields, Ms. Joan Tate, Mr. Bryan Taylor, Mr. George Tregembo, and Mr. William J. Zielinski. Personnel at the following facilities were helpful in providing opportunities to photograph mammals in their units of management: Great Smoky Mountains and Shenandoah national parks, Blue Ridge Parkway, Chincoteague and Pea Island national wildlife refuges, Mt. Pisgah National Forest, Charles Towne Landing State Park, Merchant's Millpond and Mt. Mitchell state parks, Grandfather Mountain Environmental Habitat (North Carolina), and Sea World of Florida.

Several individuals and public agencies provided illustrations of mammal species which we were unable to photograph ourselves. They are acknowledged elsewhere.

The University of North Carolina at Wilmington, through its Department of Biological Sciences, provided a professional base for our work and access to word processing and other equipment that made the task easier.

Finally, special thanks go to our wives and families who provided support and encouragement throughout the project.

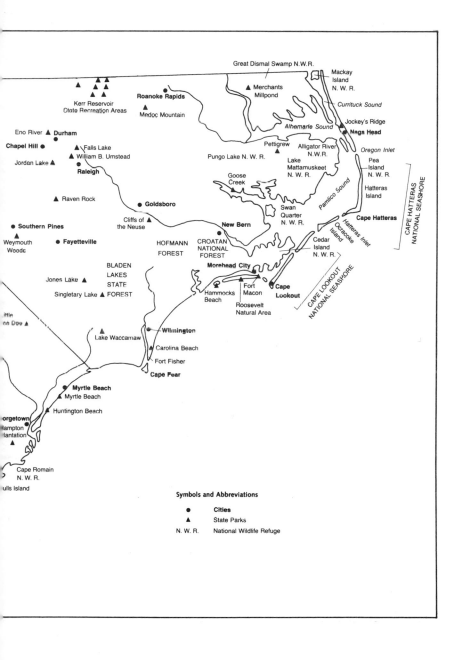

Great Dismal Swamp N.W.R.

Mackay
Island
N. W. R.

▲
▲ ▲ ▲
▲ ▲ ▲
Kerr Reservoir
State Recreation Areas

Roanoke Rapids

▲ Merchants
Millpond

Currituck Sound

▲ Medoc Mountain

Alhemarle Sound

Jockey's Ridge

Eno River ▲ Durham

Naga Head

Chapel Hill ●

▲ Falls Lake

Pettigrew

Alligator River
N.W.R.

Oregon Inlet

Jordan Lake ▲

▲ William B. Umstead

Pungo Lake N. W. R.

Lake
Mattamuskeet
N. W. R.

Pea
Island
N. W. R.

Raleigh

Goose
Creek

Hatteras
Island

▲ Raven Rock

Pamlico Sound

Cliffs of ▲
the Neuse

New Bern

Swan
Quarter
N. W. R.

Cape Hatteras

● Southern Pines

HOFMANN
FOREST

CROATAN
NATIONAL
FOREST

Cedar
Island
N. W. R.

Ocracoke Island

Hatteras Inlet

Weymouth
Woods

● Fayetteville

Morehead City

BLADEN
LAKES
STATE
FOREST

Jones Lake ▲

Singletary Lake ▲

Hammocks
Beach

Fort
Macon

● Cape
Lookout

Roosevelt
Natural Area

CAPE LOOKOUT
NATIONAL SEASHORE

CAPE HATTERAS
NATIONAL SEASHORE

ittle
on Dou ▲

▲
Lake Waccamaw

● Wilmington

▲ Carolina Beach

Fort Fisher

Cape Fear

● Myrtle Beach

▲ Myrtle Beach

▲ Huntington Beach

orgetown

Hampton
Plantation
▲

Cape Romain
N. W. R.

ulls Island

Symbols and Abbreviations

●   **Cities**

▲   State Parks

N. W. R.   National Wildlife Refuge

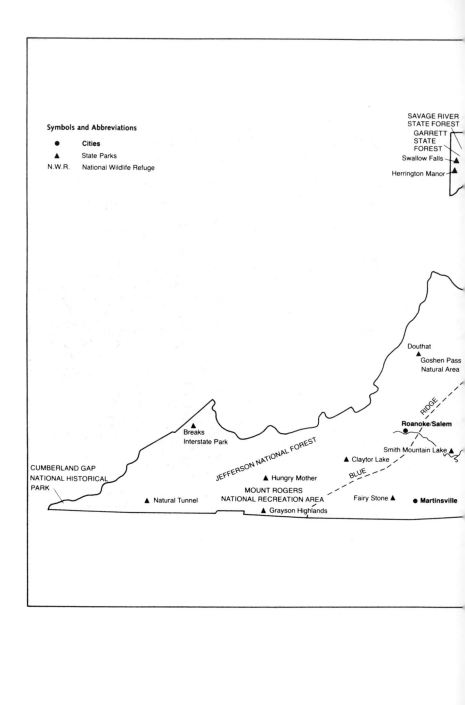

Symbols and Abbreviations

●   **Cities**

▲   State Parks

N.W.R.   National Wildlife Refuge

SAVAGE RIVER
STATE FOREST

GARRETT
STATE
FOREST

Swallow Falls

Herrington Manor

Douthat

Goshen Pass
Natural Area

RIDGE

Roanoke/Salem

Breaks
Interstate Park

JEFFERSON NATIONAL FOREST

Smith Mountain Lake

Claytor Lake

CUMBERLAND GAP
NATIONAL HISTORICAL
PARK

Hungry Mother

BLUE

MOUNT ROGERS
NATIONAL RECREATION AREA

Natural Tunnel

Fairy Stone

Martinsville

Grayson Highlands

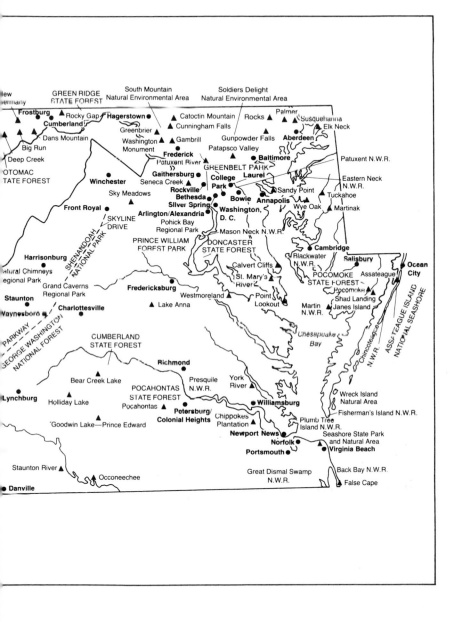

Mammals of the Carolinas, Virginia, and Maryland

Introduction

The mammalian fauna of eastern North America is extensive, but most people are aware of only the largest and most visible species. Many mammals thrive in backyards and woodlots without our being aware of their presence, and others inhabit the forests and oceans without our recognition. This book is an effort to introduce the abundance and variety of the fascinating mammalian fauna of the Carolinas, Virginia, and Maryland to those who have not yet discovered these interesting animals. Although this book was written primarily for nonbiologists, we hope that it will also be useful to students and scientists interested in the distribution, ecology, and natural history of mammals of the region. We have avoided technical language as far as possible so that persons of various backgrounds can understand the text; words that need explanation are defined in the glossary.

The mammals discussed here occur or once occurred in the region encompassing North and South Carolina, Virginia, and Maryland. This region is bounded to the east by the Atlantic Ocean and to the west by the Appalachian Mountains; although the southern and northern boundaries are politically rather than naturally defined, they generally reflect consistent patterns of distribution, and coincide with the disappearance of several northern species or the appearance of more temperate and subtropical forms.

The region is home to a varied mammalian fauna including 75 species of native terrestrial mammals. In addition, 33 species of marine mammals have been recorded in the ocean waters adjacent to these states. At least 5 species that once occurred here have been extirpated, and at least 8 species of exotic mammals have been introduced and become established. Thus the total mammalian fauna known from this four-state region is 121 species.

Many persons have published scientific and popular articles and books on the mammals of South Carolina, North Carolina, Virginia, and Maryland, providing a valuable body of knowledge about the mammals of this region. We have relied heavily on these sources in the preparation of this book and have listed those most useful to the general reader in the reference list at the end of the book.

We have combined information derived from this existing body of literature with our own original fieldwork, observations, and experiences to produce a book that summarizes current knowledge of the identification, distribution, and natural history of the mammals of this four-state region. We hope that this book will increase understanding and appreciation of mammals and that it will provide incentive for further investigation into the lives of these fascinating animals.

The Region

The region covered by this book, Maryland, Virginia, North Carolina, and South Carolina, encompasses a wide variety of environmental conditions, from mountaintop spruce-fir forests to subtropical palm-studded coastal islands. Within this wide range of habitats occurs over 85 percent of the native and introduced terrestrial and marine mammal species of eastern North America. Thus, this four-state region is an excellent place to see and study mammals.

Mountain Habitats

The western part of this region includes the southernmost section of the Appalachian Mountains, including the Blue Ridge, Great Smoky, and Allegheny mountain ranges. These mountains are very old and are worn away to relatively low, rounded peaks. Still, many peaks reach to over 5,000 feet (1,524 m) in elevation and contain habitats characteristic of more northern regions.

The tallest peaks are forested with red spruce and Fraser's fir and represent the southernmost extensions of the great spruce-fir forests that cover much of the northeastern part of North America. Here red squirrels occupy the trees while red-backed voles, smoky shrews, and deer mice live in crevices beneath exposed roots, fallen logs, rocks, and boulders. Woodland jumping mice forage near mountain streams, and eastern spotted skunks, eastern woodrats, and long-tailed shrews are resident along talus slopes and similar rocky habitats. Some species that once occurred in these mountaintop forests but that are now restricted to extremely isolated populations or which have been extirpated include the water shrew, northern flying squirrel, snowshoe hare, porcupine, and fisher. Excellent examples of this habitat type are found in the Shenandoah and Great Smoky Mountains national parks in western Virginia and North Carolina, respectively.

Native grasslands called balds also occur on some mountain peaks, and these provide a rather unusual habitat for mammals. In places like Gregory Bald, along the North Carolina–Tennessee border, mammals such as the meadow vole, least weasel, meadow jumping mouse, and masked shrew may be found.

Most mountain slopes are covered by vast deciduous forests composed of a variety of oaks, maples, beech, poplar, and other hardwoods. The most impressive of these sites are the rich cove forests that have developed in many cool, moist mountain valleys. Typical cove forest mammals are the hairy-tailed mole, eastern chipmunk, gray squirrel, black bear, northern short-tailed shrew, and white-footed mouse.

Forests dominated by oaks and hickories occupy moist mountain slopes, whereas pine forests cover the drier and rockier sites. Each of these

Coniferous forest on Mt. Mitchell, North Carolina.

forest types undergoes characteristic patterns of change following such disturbances as fire or timbering, and the result is a patchwork ranging from shrub thickets to old mature forests. These varied habitats then become home for a wide variety of mammals such as northern short-tailed shrews, cottontails, eastern chipmunks, gray squirrels, white-footed mice, raccoons, and white-tailed deer.

Caves provide another interesting habitat for many mammals, especially several species of bats. The little brown myotis, Keen's myotis, and the small-footed myotis are relatively common in many caves in the mountains, whereas the gray myotis, Indiana myotis, and Townsend's big-eared bat, which are listed as Endangered Species, are known in the region from only a few caves in the southern Appalachian Mountains. Caves also occasionally provide shelter for eastern woodrats, striped and spotted skunks, and black bears.

Piedmont Habitats

As one descends from the mountains and enters the rolling hills of the piedmont, agriculture becomes the dominant theme. Here forests are isolated and agricultural fields are the primary feature of the landscape. Where forests remain, they are often in small blocks isolated from other units or follow river floodplains unsuitable for agriculture. These forests are primarily deciduous, comprising a variety of oaks and hickories, and often have an abundance of flowering dogwood in the understory. On drier or more recently disturbed sites, coniferous forests dominated by shortleaf, Virginia, and loblolly pines occur. Some mammals characteristic of these

Mountain bald near Roan Mountain, North Carolina. Photograph by Wm. David Webster.

areas are gray squirrels, southern flying squirrels, eastern chipmunks, short-tailed shrews, gray foxes, raccoons, and white-footed mice.

Of special interest are the deciduous forests that develop on the moist, fertile floodplains of streams and rivers crossing the piedmont. Such forests act as natural corridors for the passage of mammals because they extend for many miles in uninterrupted strips. They also serve as refuges for mammals forced from more disturbed upland sites. Floodplain forests are home to a resurgent beaver population. Other typical residents are raccoons, long-tailed weasels, opossums, white-tailed deer, and either white-footed or cotton mice.

While most of the piedmont consists of uplands, there are often small areas within the floodplain forests that retain shallow water throughout much of the year; these small swamps are important habitats for many mammals. Beaver ponds often create habitat similar to natural swamps, and this wetland type is actually increasing in some localities. Such mammals as mink, muskrats, and river otters inhabit these wetlands, and many other species come to them for water in times of drought or for refuge when nearby more accessible habitats are disturbed. Meadow jumping mice, meadow voles, striped skunks, and golden mice often inhabit surrounding low-lying grasslands and brushy areas. Marshes surrounding farm ponds, lakes, and reservoirs also provide important wetland habitats for such mammals as muskrats, raccoons, mink, and star-nosed moles.

It is in the old-field habitats of the piedmont that many species of small mammals are most common. As agricultural fields are left fallow in the fall and winter, a cover of crabgrass and

Mountain stream flowing through a deciduous forest. Photograph by Walter C. Biggs, Jr.

Deciduous forest in southwestern Virginia under cover of snow.

annual weeds grows and blankets the ground. If such fields are not planted, herbaceous plants such as horseweed cover the ground the next spring. In a few years abandoned fields become vegetated with broomsedge, and after a few more years pines appear. This characteristic pattern of succession of plant communities, which has been well studied by ecologists, is called "old-field succession." It generally leads after many years to the mature oak-hickory forests characteristic of undisturbed sites.

At each stage in this pattern of changing plant species there is a corresponding change in the mammalian fauna. Eastern harvest mice, oldfield mice, and eastern moles are often found in the crabgrass fields that usually follow the harvest of agricultural crops. By the time broomsedge becomes dominant, cotton rats, least

shrews, eastern cottontails, and bobcats have become important members of the community. As pines shade out the grasses, cotton rats disappear, but eastern cottontails may continue to use the pine forest for cover although they feed in nearby habitats dominated by herbaceous plants; gray squirrels and short-tailed shrews may occupy the pine forests. As hardwoods displace the pines, southern flying squirrels, white-footed mice, gray squirrels, and several species of bats become more abundant, and the fauna becomes that characteristic of a mature forest community. White-tailed deer, woodchucks, striped skunks, red foxes, and eastern cottontails may be observed at the edge of these forested areas.

Coastal Plain and Ocean Habitats

As one travels east, the rolling hills and red clay soils of the piedmont give way to the sandy soils and flatlands of the coastal plain. Agriculture is still important, and all stages of old-field succession can still be found, but the amount of forested land begins to increase. Large areas of low, poorly drained land long remained undeveloped because the necessary clearing and draining could not be done profitably. Unfortunately for animals that inhabit wetland forests, with new techniques and equipment these lowland habitats are now being cleared, drained, and converted to agricultural lands.

The coastal plain is an area of diverse habitats in spite of the relatively low relief and slow drainage associated with the gradual slope of the land toward the sea. On the higher parts of the coastal plain are found oak-hickory forests similar to those of the piedmont. Most of these upland forests, however, have strong components of pine, and mixed pine-hardwood forests are common.

In south-central North Carolina and central South Carolina, a distinct band of sandhills lies along the border between the piedmont and coastal plain. These rolling hills and infertile sandy soils are forested by extensive stands of longleaf pine, which are home to such mammals as fox squirrels and white-tailed deer.

Rivers slow their passage in the coastal plain and begin to make long winding loops over extensive floodplains. Where flooding is prolonged into the growing season, cypress or gum swamps develop. The Pocomoke River on the Eastern Shore of Mary-

Foothills of Virginia, viewed from Sharptop Mountain. Photograph by Walter C. Biggs, Jr.

Floodplain forest.

land is the northernmost of these southern river swamp systems. South of the Chesapeake Bay, most coastal rivers develop swamp forests along their lower reaches. The Great Dismal Swamp, which lies along the Virginia–North Carolina boundary, and the Francis Marion National Forest near Charleston, South Carolina, contain excellent stands of swamp forests. They provide important habitat for such mammals as marsh rabbits, golden mice, bobcats, southeastern shrews, gray squirrels, mink, raccoons, white-tailed deer, and black bear. The Dismal Swamp is also home to several very interesting but poorly known mammals, such as the Dismal Swamp races of the southeastern shrew, northern short-tailed shrew, southern bog lemming, and meadow vole.

Pocosins are another coastal plain habitat of some importance to mammals; these dense shrub swamps usually have a bed of sphagnum moss covering the surface, and are often underlain by deep peats. Although pocosins are scattered throughout the coastal plain of the Carolinas, they are most extensive in northeastern North Carolina. These dense, almost impenetrable wetlands are important refuges for black bears and also contain such common mammals as marsh rabbits, gray squirrels, cotton mice, and white-tailed deer. As habitats, they have been little studied and are in need of further investigation.

As one approaches the coast, forested wetlands give way to marshes and shallow bays. The Chesapeake Bay, one of the most extensive marsh-bay systems in the world, is located in Virginia and Maryland. This magnificent, complex system of coastal river

Broomsedge field typical of old-field succession in piedmont habitats.

mouths and shallow fresh and brackish bays and marshes is the home of the largest population of muskrats along the Atlantic coast. In addition, river otters, mink, and raccoons are present, and smaller mammals such as marsh rice rats and meadow voles are locally abundant. Similar brackish marshes extend southward throughout the region and are likewise important mammal habitats.

Brackish marshes grade slowly into salt marshes as salinities approach the strength of seawater, about 35 parts per thousand of total salts. An extensive band of salt marsh exists between the barrier island chain and the mainland from Maryland to southern North Carolina and behind the sea islands of South Carolina. These productive marshes are home to such mammals as least shrews, meadow voles, marsh rice rats, raccoons, and

mink. In several places, such as along the North Carolina Outer Banks, nutria, large rodents native to South America, have been introduced and are now abundant.

The series of barrier islands that extends from Maryland to southeastern North Carolina is replaced in South Carolina by more widely spaced sea islands. The mammalian fauna of these barrier and sea islands is usually much like that of similar habitats on the nearby mainland, but often with fewer species, especially if the islands are well isolated from each other or from the mainland. Mammals such as raccoons, white-tailed deer, house mice, and marsh rice rats have little difficulty in crossing channels and are common on most islands.

The presence on several of these islands of feral populations of once-domesticated animals is of particular

Mixed cypress-gum swamp typical of lower coastal plain.

interest. Chincoteague and Assateague islands on the Eastern Shore and Ocracoke Island on the North Carolina Outer Banks have remnant populations of feral horses—the wild ponies of Chincoteague and Ocracoke. Shackleford Island near Beaufort, North Carolina, has feral sheep, and Baldhead Island on the southeastern North Carolina coast had feral hogs until very recently. Bulls Island in South Carolina once had feral cattle, and a number of other islands had or still have remnant populations of domestic livestock. In addition to these feral animals, Assateague Island has a population of introduced sika deer, a species native to Asia, and several coastal islands, including Hatteras Island in North Carolina, have populations of nutria, stocked there to provide an additional fur source.

Perhaps the most important of the coastal mammal habitats is the one we know least about—the Atlantic Ocean. At least 33 species of marine mammals have been recorded in the offshore waters of the Atlantic Ocean bordering this region. Some of these, such as the bottle-nosed dolphin, regularly enter the bays and coastal river mouths and are relatively well known. Young harbor seals frequently stray southward to the mid-Atlantic coast, and other mammals, such as the dwarf and pygmy sperm whales, are known primarily from individuals that became stranded on coastal beaches. Many species of whales, dolphins, and porpoises regularly pass through mid-Atlantic coastal waters during seasonal migrations, and the manatee still occasionally moves northward to the Carolinas during the summer. Other marine mammals apparently are present year round, but generally little is known of their life histories and ecologies. Cetaceans and

Brackish Chesapeake Bay marshes on the Eastern Shore of Maryland.

Barrier island and inshore ocean at Cape Lookout, North Carolina.

seals remain a fascinating part of the mammal fauna, to be glimpsed only at sea from the beach or a boat, or viewed in awe when they become stranded on the beach.

The mammals of the Appalachian Mountains are quite distinct from those of both the piedmont and coastal plain, which have somewhat similar faunas. Mountain species include several with northern affinities; for example, several shrews and mustelids that inhabit the spruce-fir forests of Canada and the northeastern United States extend their ranges southward along the Appalachians. In contrast, the mammals of the piedmont and coastal plain provinces are relatively homogeneous, and many species of mammals are distributed widely in both. Of course, the marine fauna of the Atlantic Ocean is unique and varied, including such species as the manatee and Bryde's whale, which reach the northern limit of their range in the four-state region, and others, like the Atlantic white-sided dolphin, which reach the southern terminus of their range here. All told, the greatest number of species of mammals is found in the vicinity of the Great Smoky Mountains National Park; the lowest, on the barrier islands and in the sandhills of the Carolinas.

Characteristics and Adaptations of Mammals

Mammals tend to be secretive and many are nocturnal; they are usually seen in the wild by humans only in tantalizing glimpses. A momentary view of a white-tailed deer browsing at the edge of a meadow in the soft light of dawn, a black bear crashing through the dense brush of a pocosin, or a gray squirrel scurrying along a high limb of an ancient oak induces excitement in all but the most insensible observer. More detailed observation of live mammals usually requires that we patiently seek them out in their native habitats or visit parks or zoos where they are protected and tend to lose their fear.

The fossil record reveals that the earliest mammals appeared about 250 million years ago and that they arose from an ancient group of reptiles that were "mammal-like" in certain of their structures. Small primitive mammals were relatively inconspicuous contemporaries of the dinosaurs. After the extinction of the great reptiles about 65 million years ago, mammals flourished, became highly diverse, and assumed a position of major importance among the creatures of the earth. Many mammal groups that evolved became extinct while others survived. The present mammalian fauna is the product of a long process of evolutionary development and adaptation.

Biologists separate animals into major groups, or phyla, based upon distinctive differences in body structure and way of life; phyla are subdi-vided into classes, orders, and other categories of classification. Fishes, amphibians, reptiles, birds, and mammals are vertebrate animals, each in a separate class of the Phylum Chordata. These vertebrates are alike in having a backbone, a cranium that encloses the brain, and other common features, but each class differs significantly from the others. Mammals (Class Mammalia) possess characteristics that readily distinguish them from their fellow vertebrates.

Most vertebrate animals reproduce by laying eggs that develop outside the body. In contrast, in all but the most primitive group of mammals (the egg-laying monotremes, or duck-billed platypus and echidna of Australia and New Guinea) fertilized eggs are retained in the uterus of the female, where the embryos develop, and young are born alive. An advanced placenta allows nutrients, oxygen, and other substances to pass from the blood of the mother to that of developing embryos. The distinctive mammary glands of females secrete milk, which is suckled by newborn offspring and provides nourishment until they are weaned.

Hair is unique to mammals and is their most obvious characteristic. In most species it covers the body, furnishing insulation against heat loss and, with oil secreted by skin glands, provides a water-resistant covering. The pelage of most mammals is protectively colored, helping them to avoid detection; usually mammals molt, or shed, seasonally or annually. In most species the pelage consists of short underfur and longer guard hairs that cover the surface; in a few, such as moles, guard hairs are velvet-like and underfur is absent. Hair is sparse on some mammals, such as armadillos, and reduced to a few bristles on the snouts of some whales and porpoises.

Mammals share with birds the ability to regulate body temperature internally. While other vertebrates generally derive heat from their environment and experience fluctuations of body temperature as environmental conditions change, mammals and birds produce heat internally as food is metabolized. Most mammals maintain a relatively high and constant body temperature as their "internal thermostat" increases production and conservation of heat energy in cold weather and causes loss of heat in hot weather through sweating, panting, and radiation from the body surface. As a result, most species can remain active throughout the year, and many are adapted to such diverse habitats as deserts, the Arctic, and the depths of the oceans (though these oceanic species must come to the surface for air). Some mammals avoid extreme conditions by such behavioral responses as migration, hibernation in winter, aestivation in summer, or being inactive in dens or burrows.

Mammals possess teeth that are well developed and usually differentiated into four types, namely (from front to rear on both the upper and lower jaw) incisors, canines, premolars, and molars. These specialized teeth permit a variety of foods to be sheared, sliced, crushed, or ground and, therefore, digested and metabo-

Litter of suckling deer mice.

lized thoroughly. The teeth of various mammal groups are adapted structurally to diet and are therefore useful in identification. Incisors are generally chisel-shaped for nipping or gnawing and are especially well developed in rodents and rabbits. Canine teeth tend to be elongate and pointed for defense and to capture, hold, and kill prey; they are best developed in meat eaters, such as wolves and mountain lions, and are absent in many plant eaters. Premolars in carnivores may be bladelike to assist in cutting meat; in herbivores they usually are comparable to the molars, which tend to be broad with relatively flat surfaces bearing a variety of cusps and ridges. Molars are used to crush or grind foods, primarily vegetation, and are best developed in large herbivores such as deer and cattle. These specializations of tooth structure in mammals, along with highly efficient jaws and strong jaw muscles, make for a powerful bite and an effective chewing action. Mammals also have two sets of teeth: a deciduous ("milk") group is later replaced by permanent teeth.

A feature of mammals is the complex tooth structure; here canine, premolar, and molar teeth are visible.

Mammals are highly active and mobile animals. They typically have four well-developed limbs with digits, or toes, that end in claws, nails, or hoofs. Various specializations of limb structure reflect much about the way of life of animals: where they live, how they move from place to place, and how they pursue prey or escape predators. Among species that are ambulatory, or walkers, such as raccoons, bears, shrews, and humans, the entire foot, including the heel, touches the surface of the ground as the animal walks. Many of the carnivore species and the ungulates, or hoofed mammals, depend on speed for survival and their limbs are modified for running. In predatory carnivores the heel is raised and only the toes and "ball" of the foot touch the ground, lengthening limbs and stride. Ungulates, such as deer, antelopes, and horses, which depend on speed to escape predators, run with only the tips of the toes on the ground; the bones of the toes are enlarged and protected by strong hooves. The numbers of toes and bones of the feet and lower legs in these mammals are greatly reduced; for example, horses have one toe on each foot, and cattle have two.

Other mammals exhibit additional locomotor adaptations. Kangaroos, rabbits, and some species of mice have combined speed when running with the ability to jump; they have greatly enlarged and elongated hind legs and feet, and some species, like the jumping mice, have long tails for balance. Opossums are arboreal and cling to the branches of trees with the aid of opposable digits and a prehensile tail. Tree squirrels have strong, sharp claws that enable them to scamper up vertical tree trunks and jump from limb to limb. "Flying" squirrels have the unusual ability to glide relatively long distances from tree to tree aided by folds of skin on each side between the front and hind limbs. Bats, with forelimbs modified as wings, are the only mammals that fly, often covering great distances to feed or migrate. Some mammal species spend much or all of their time underground; for example, moles are fully fossorial, and woodchucks are partially so. Their forelimbs are modified for burrowing, tending to be shortened and very powerful, with broad front feet equipped with long, sharp claws.

Many species of mammals are adapted to move in water. Whales, porpoises, and manatees are fully aquatic, unable to come onto land. Their bodies are streamlined, and forelimbs are modified into flippers used in steering while the broad flattened flukes of the tail provide propulsion; the hind limbs are absent externally but small vestiges of limb and girdle bones are present internally in some species. Other aquatic mammals such as seals and sea lions spend most of their lives in water but come onto land periodically, primarily to bear young. Both front and hind limbs are present externally, modified into flippers with toes that are fully webbed. Beavers, river otters, muskrats, and

water shrews are examples of partially aquatic mammals, equally at home on land and in water. Most have webbed hind feet and a generally flattened tail that enhances swimming.

The complex physical and physiological activities of mammals are controlled by a significantly advanced nervous system. The mammalian brain is unusually large compared to that of other vertebrates, due primarily to the development of the cerebral hemispheres and especially the overlying mantle of gray matter. In most higher mammals this structure is complex and voluminous, characterized by extensive foldings or convolutions. It functions as a control center which dominates the functioning of much of the remainder of the brain. Perhaps most importantly, this part of the brain is responsible for increased intelligence.

The sense of smell is acute in many groups of mammals, including insectivores, carnivores, and rodents; in contrast, olfaction is poorly developed in manatees and cetaceans. Mammals also depend heavily on their ability to hear, and this sense is highly developed in most groups. Only mammals have developed fleshy pinnae associated with external ear openings; manatees, cetaceans, and some insectivores and seals, however, have secondarily lost this structure. The ability to see is also well developed in most mammals, and nocturnal mammals often have enlarged eyes and other adaptations that enhance vision in dim light. Certain insectivores, bats, and cetaceans do not rely heavily on sight, and their eyes are reduced in size. Vibrissae on the muzzle and other specialized hairs on the body give many mammals an enhanced sense of touch which is especially useful to those that feed at night.

Another unusual adaptation used by some mammals to orient to their environment is echolocation—the ability of an animal to produce high-pitched sounds and interpret the echoes that return from obstacles or potential prey. Echolocation is best developed in many species of bats and toothed whales, but some insectivores, pinnipeds, and rodents also use this remarkable ability, and several other mammals are suspected of having it.

Additional distinctive characteristics and adaptations of mammals involve details of skeletal and soft anatomy and are discussed in several of the publications listed in the reference section.

Observation and Study of Mammals

The methods commonly employed in the study of mammals include direct observation of animals or their sign in natural habitats, capture and collection of specimens for study in museums or laboratories, and visits to museums, zoos, and parks. Many study procedures, such as live and kill trapping, mark and recapture of individuals of a population, and use of radio-location telemetry, drugs for immobilization, and radioisotopes, are appropriate only for professionals. Interested and thoughtful persons, however, can gain personal and direct experience with these most fascinating animals in several ways.

Mammal Sign

The casual observer only infrequently sees mammals in their natural habitats; one may be fortunate enough to glimpse a woodchuck along a roadside, a chipmunk or red squirrel looking for handouts in a campground or picnic area, or the glistening eyes and bushy tail of a fox in the beam of a car's lights. The presence of mammals is most often revealed by the sign they leave, such as tracks, burrows, runways, nests, food residues, or fecal droppings. Such spoor often make it possible to determine the identity of specific mammal species, as well as much about their habits and distributions. Well-informed "sleuths" can learn to interpret the clues mammals leave behind and thus gain much in-

formation about the animals themselves.

Mammals leave footprints and often tail markings in soil, on dust-covered surfaces, or in fresh snow. Sometimes an observer can follow tracks to where the animal ate, drank, slept, or had an encounter with another animal. One may read in a thin blanket of snow a tale of pursuit and struggle between predator and prey, follow an animal to its den, or see where it foraged. The tracks of most larger mammals are relatively easy to identify and interpret because of their characteristic patterns. Those of small mammals are more difficult; their identification often requires knowledge of what species occur in the area and the presence of additional clues or sign. It is often difficult to distinguish between the tracks of closely related species because they may be almost identical. Several reference sources usually available in local libraries provide illustrations and descriptions of tracks and other sign of most North American mammals; these are useful on treks out-of-doors, or at home for comparison with photographs or plaster casts made of tracks discovered in the field. Among the references listed at the end of this volume are guides to identification of mammal tracks and other sign.

A careful observer exploring a grassy hillside, an overgrown abandoned field, a forest floor covered with duff, or a streamside thicket is likely to discover a variety of trails and

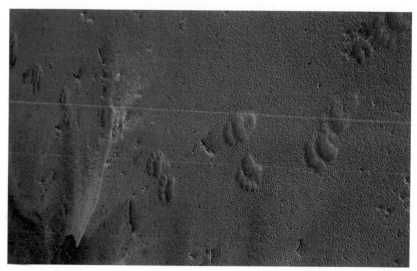

Black bear tracks in the soft sand of a coastal plain roadside. Photograph by Wm. David Webster.

runways made by mammals. Runways are formed as animals habitually use the same paths in their travels to and from nests, water sources, feeding sites, or other points within their home ranges. They are formed by mammals as large as a white-tailed deer or as small as a shrew. A runway may be completely open (as one formed by beavers moving between a stream bank and the edges of fields and forests), partly covered (as by cotton rats moving through the grasses of open fields), or completely covered, tunneled beneath vegetation on a verdant hillside in an intricate interconnecting system (as by voles and numerous other small mice). Shrews form small, often obscure, runways beneath forest leaf litter. Species, such as white-footed mice, that do not produce runways themselves may use those of others; their presence may be detected by other sign such as fecal droppings. Some species, including eastern harvest mice, move about randomly, seemingly disregarding the use of runways.

Many mammal species produce excavations that provide evidence of their presence. These are highly variable among the different species, from shallow depressions (or forms) made by hares for resting and birthing young to more elaborate and often conspicuous burrows and tunnels beneath the soil. Underground excavations may serve as dens or nest sites, storage chambers, or a means to reach such foods as roots or earthworms. Entrances to tunnels may be vertical or at an angle; a single system of burrows and tunnels may have several entrances. Soil from the burrow may be left in mounds about the entrance, as with woodchucks, or scattered and

Mouse emerging from runway leading to nest.

less conspicuous. Clues to the identity of a mammal inhabiting a burrow include the size of the entrance (and thus the size of the animal), footprints, fecal droppings, and bits of fur about the opening. A burrow entrance in a stream bank may be that of a beaver or muskrat, especially if it is below water level, or a river otter if there is a mudslide nearby. Large burrows may have the distinct musky odor of a skunk or contain the remains of small prey left by a predator such as a fox. Perhaps the most familiar burrowing mammals are the moles that dig tunnels just below ground surface, which are seen as elongate ridges on the surface with small mounds of excavated earth at intervals along their length.

Many mammals have nests or dens in which they spend much of their time sleeping, rearing young, avoiding predators, or seeking a more comfortable temperature. Burrowing species usually construct nests within their system of tunnels, lining them with fur, grasses, leaves, or other suitable materials. Some species move into burrows abandoned by other mammals. Nests above ground may be found under cover of protective vegetation, in depressions or holes in the earth, in crevices among rocks, or in trees. A loose assemblage of leaves high in a tree reveals the presence of tree squirrels; they live in leaf nests in summer but often move into hollows when it is cold. Opossums and raccoons are likely to reside in hollow trees or logs year round. An elaborate mound of twigs, leaves, and other debris at the base of a tree, in a rock pile or cave, or even in an old abandoned building likely was constructed by an eastern woodrat. The presence of

Entrance to chipmunk burrow in mountain habitat. Photograph by Walter C. Biggs, Jr.

beaver in an area is unmistakable because they dam streams with felled tree trunks and other vegetation, producing a pond within which they build large, conical lodges of sticks and mud. Muskrats construct similar but smaller houses of marsh vegetation, usually at the surface of estuarine waters. Both beaver and muskrat may choose to nest underground in the bank of a marsh or stream, especially if suitable building material or habitat is lacking. Bears and foxes often find good den sites beneath fallen logs or boulders or in caves and hollows. Bats, our only flying mammals, roost in caves, trees, rock crevices, or old buildings.

A mammal's fecal droppings, or scat, often provide excellent clues to the identification of the animal, especially in combination with tracks and other sign. Scat of larger mammals, such as carnivores and ungulates, usually is quite distinctive, whereas that of small species, such as various mice, is less so. Examination of scat can provide information about the activities and feeding habits of the animals—what was eaten and the relative quantities of different food items. Scat can be found along trails, in chambers of tunnels and burrows, on rocks or logs, in grassy areas where grazing has occurred, or in mines, caves, or the attics of buildings. Bright green droppings in a runway indicate the presence of southern bog lemmings; brown-colored ones, meadow voles. The dark pellets of rabbits are common along their trails and in resting areas, and the scat of deer on trails or at a forest edge is unmistakable. Scat containing fish scales suggests the presence of river otters; bone fragments and fur tell of predators such as foxes; pieces of crayfish skeleton, insects, and plant matter reveal the omnivorous feeding habits of raccoons. Accumulations of fecal matter in a cave, beneath a tree, or in an old building may betray a bat roost. Illustrations useful in the identification of scat are contained in several of the volumes listed in the references section.

In addition to scat, other mammal sign that can be observed in the field provide insight into the feeding activities of various species. Deer browse on leaves and tender shoots of shrubs and trees, often pruning the plants as high as they can reach. Voles forage on grasses and forbs and often leave cuttings along runways. Stripped bark and felled saplings result from a meal by a beaver. Wild pigs, nine-banded armadillos, and skunks leave their imprint as they root for food in the soil

of a forest floor. Caches of food are made by such mammals as shrews, mice, and squirrels; stored items may include seeds, nuts, fruits, fungi, grasses, roots, snails, and insects. Storing activity is likely to be most intense in fall and early winter, especially by mice and squirrels. Discarded shells of seeds, nuts, and snails or a pile of pine cone scales reveals where a mammal has paused recently for a meal.

A stroll through a forest or meadow or along a creek or ravine may provide the curious individual glimpses of other, less obvious indications of the presence of mammals. A bobcat or bear may leave claw marks on a tree trunk. Strands or bits of hair may cling to a bush, rock, or fence as a result of the passing of a mammal. Velvet from deer antlers might be found hanging on a branch. Rodents often

Muskrat house in a brackish coastal marsh.

Evidence of a feeding squirrel—a pile of pine-cone scales. Photograph by Walter C. Biggs, Jr.

gnaw on bones or antlers found on the ground, leaving conspicuous teeth marks. Other possibilities abound.

Capture and Collection of Mammals

It is often necessary that mammals be captured or collected in order to study in detail certain aspects of their biology. Many investigations require that they either be brought into a laboratory environment for observation and experimentation or marked for individual identification and released for further study in the field. Other investigations require that they be preserved for deposit in research collections in museums, colleges, or universities. These methods of study are carried out by trained and experienced scientists, or with their assistance and guidance. All capture and collecting of mammals should be done for a valid reason—that is, to enhance knowledge and understanding of our invaluable mammalian fauna—and in a manner that assures the continued well-being of species in their natural environments. Numerous federal and state laws and regulations govern the capture, possession, transport, and salvage of various animal species. It is imperative that everyone be familiar with and adhere to all applicable laws. Also, it is necessary that permission to trap or otherwise conduct studies be obtained from landowners. Collecting and research may be done on certain state and federal land, but permission usually is necessary and game laws must be obeyed.

Mammals in captivity provide data concerning such matters as the behavior of the species in response to a variety of stimuli, their pattern of reproduction, length of the gestation period, growth and development of offspring, daily or seasonal activity cycles, and pattern and timing of molting of the pelage. Captive animals must be cared for appropriately; specific requirements for space, nesting material, food, and water must be met. Attention must be given to their comfort and needs. Ideally, the natural habitat of the species would be simulated in the cage or aquarium used to hold the animals. If practical, the animals may be returned to their original native habitat when studies are concluded.

Mammals that are marked and released for recapture enable mammalogists to investigate such matters as movements within the animal's home range, trends and cycles in the size of local populations, and other aspects of the ecology of the species. Marking may be accomplished by such simple procedures as toe or ear clipping, application of a tag or color dye, or use of arm or leg bands, especially for bats. However, more sophisticated methods may be required. A radio transmitting sound pulses and attached to the neck or back of a mammal allows a researcher with a directional antenna and receiver to acquire information on the physiological state of the animal as well as its movements and activities. Implanted radioisotopes, judiciously employed, make possible the tracking of mammals. Large mammals that must be captured alive for examination, removal from an area, or marking may be immobilized with drugs such as nicotine salicylate and succinylcholine chlo-

ride, thus preventing harm to the animal and the researcher. Specially designed syringes attached to arrows or fired from a gun are used to administer the drugs; regulations exist which govern their use.

The collection and preparation of specimens is necessary when making faunal surveys in a given geographic area to document the presence of a species, to determine the kinds of parasites carried by the animals, to analyze stomach contents and thereby gain information about the animal's food preferences, to make detailed taxonomic studies, and to gain additional information otherwise unobtainable. The nature of a study usually dictates the number of specimens that must be taken; for example, one or a few may be needed to document the presence of a species locally, whereas a relatively large series of individuals is required for a detailed analysis of variation. A variety of traps and trapping techniques are available, depending upon the objectives of the study and the mammal being investigated. Live traps, pitfall traps, mist nets, and bat traps usually allow an animal to be taken alive and unharmed. Snap traps and other forms of kill traps are used in collecting specimens. These must be used judiciously.

Both data and specimens of mammals often may be obtained by means other than trapping. Many mammals are killed by automobiles and may be salvageable. Skulls and other parts of skeletons may be found in fields or woods; pellets regurgitated by owls usually contain skeletal remains of small mammals preyed upon. Hunters or trappers may be sources of both specimens and data relative to the distribution and natural history of game species. Information from sources such as these provide the opportunity for amateur mammalogists to contribute to the science of mammalogy. Material or data which might be of potential value should be made available to museums or biologists at colleges or universities in the vicinity.

Essential to any study of mammals is the recording of data, or the keeping of accurate and detailed records and field notes. All specimens, if they are to be of scientific value, must be accompanied by such information as date of encounter, precise locality where obtained, method of collection, and name of collector. Several books listed in the reference section provide more detailed discussions on methods that may be employed in the study of mammals, including the use of traps, preparation of specimens, and recording of data and field notes.

Museums, Zoos, and Parks

Museums, zoos, and parks provide opportunities for both scientists and nonscientists to observe and study mammals. Many such facilities, operated either by an agency of a local, state, or federal government or by private organizations and individuals, are accessible in the region.

Museums, in addition to housing collections of mammals for the purpose of scientific study, are open to the public and usually maintain attractive exhibits of mounted specimens. They often offer educational and interpretive programs designed to enhance public awareness of many of the organisms with which humans share the earth. Some of the larger museums in the four-state region are the National Museum of Natural History in Washington, D.C., the Charleston Museum in Charleston, South Carolina, and the North Carolina State Museum of Natural History in Raleigh, North Carolina.

In zoos people often can see at close range animals that they might not otherwise encounter, either because the animals are too secretive in nature, are nocturnal, or do not occur in the area. Most modern zoos exhibit live mammals in natural settings, thus increasing the appreciation and understanding of animals in relation to habitat. In addition to exhibiting and caring for animals, zoos often serve a vital function by helping to conserve a variety of species. Many rare and endangered mammals may someday be perpetuated only in such man-made environments. Major zoos in the four-state region include the National Zoological Park in Washington, D.C.; the Baltimore Zoo and the National Aquarium in Baltimore, Maryland; Lafayette Zoological Park in Norfolk, Virginia; the North Carolina Zoological Park in Asheboro, North Carolina; and Riverbanks Zoo in Columbia, South Carolina.

In addition to major museums and zoos with large numbers of species, there are many area or municipal nature centers and small commercial zoos in each of the four states. These often combine museum exhibits with a few species of live mammals, often in natural surroundings.

Mammals in national and state

parks and wildlife preserves are less visible than those in zoos; being protected, however, they tend to have less fear of humans than do mammals living in other wild environments, and the probability of observing many of them is enhanced. Each of the four states has an established park system, and many state parks provide special opportunities for a careful observer of animals. The region also includes several units of the National Park system, each rich in species of mammals. Rare is a visit to Great Smoky Mountains National Park, Shenandoah National Park, or the Blue Ridge Parkway during which one does not observe such mammals as white-tailed deer, woodchucks, eastern chipmunks, tree squirrels, striped skunks, and black bear; at least 23 species of larger and more visible mammals are listed for these parks. Cape Hatteras, Cape Lookout, and Assateague national seashores, perhaps best known as bird-watchers' paradises, are also home to many species of mammals. Eighteen national wildlife refuges are within the region. These preserve important units of animal habitat, and you may be successful in seeing many mammals or their sign in these protected environments. The locations of state parks and national seashores, parks, and wildlife refuges are shown on the state maps on pages viii–ix and x–xi.

Population Regulation and Conservation of Mammals

The number of mammals in a given population is regulated by a series of ecological factors, such as competition and the availability of suitable habitat; various environmental factors including weather, diseases, and parasites; predators; crowding; and, especially at present, the activities of man. The numbers of each species are regulated by one or more of a long list of factors that affect the general health of the population. In a given locality, one species may be limited by a lack of suitable habitat, another may be restricted by the presence of a competitor, and a third may be having difficulties as a result of a series of severe winters.

In spite of the variables, we can look at the dominant ecological determinants operating in this region and see several trends. Of primary importance is the heavy hand of man changing a forested landscape to one of open fields and other nonforested habitats; this also means that the larger truly wild and remote areas are shrinking. Therefore, mammals that require large areas of remote forest are becoming rare; for example, black bears and fox squirrels no longer exist in the piedmont and their numbers are declining in the coastal plain where agricultural clearing operations are widespread. On the other hand, mammals that find suitable habitat in woodlots, agricultural fields, and highway rights-of-way are becoming more abundant; hence the increasing numbers of white-tailed deer and eastern cottontails and the recent range extensions of the coyote, hispid cotton rat, and woodchuck.

The numbers of many insectivorous mammals have declined in recent years during which the use of persistent pesticides was practiced widely. Pesticides have also entered the ocean communities, and many cetaceans, especially the baleen whales, are accumulating high levels of these toxic chemicals. We anticipate that populations of such groups as bats and whales may recover as the use of pesticides is more carefully regulated in the future.

Many species of cave-dwelling bats also face increasing levels of disturbance as the popularity of cave exploration increases. Several species that use caves for daytime roosts, hibernacula, or maternity colonies have become threatened or endangered largely due to such disturbances.

Competition between closely related species also affects the distribution and abundance of mammals. In this region, the New England cottontail is becoming rare and localized as its close relative, the eastern cottontail, becomes more abundant and widespread. A similar situation exists between two closely related species of squirrels. The northern flying squirrel is now restricted to only a few isolated mountaintops, its former range being usurped by the southern flying squirrel; habitat destruction is further contributing to this replacement.

Another important factor that governs the abundance and distribution of mammals is change in climate. The region is currently experiencing a warming trend, and some mammals are expanding their ranges northward. For example, the nine-banded armadillo, introduced into Florida in the 1920s and 1930s, now occurs in Georgia and may be established in southern South Carolina as well. The Brazilian free-tailed bat also has been expanding its range northward, now reaching central North Carolina, and the hispid cotton rat now inhabits the outskirts of Richmond, Virginia.

Most large predators are probably gone from this region. We are not apt to see gray wolves here again, but the mountain lion may be making a comeback. The increasing abundance of white-tailed deer in the region is providing a bountiful supply of this cat's favorite food, and we may once again have the opportunity to catch a glimpse of this magnificent animal in the more remote parts of some eastern states.

Populations of mammals fluctuate as conditions for their existence vary, and occasionally a species may completely die out and become extinct, or it may be eliminated from portions of its former range and be said to have been extirpated from that region. No species of terrestrial mammal is known to have become extinct in eastern North America since this country was settled, but several have been extirpated from large portions of their former ranges. Thus, the gray wolf

lives on in portions of Michigan, Minnesota, and Wisconsin, in much of Canada, and Alaska, but it has been extirpated east of the Mississippi River. A number of species have become extirpated from portions of former ranges in this region. The porcupine has been extirpated from Virginia and the gray whale no longer occurs in the ocean waters adjacent to our coast. Many other species or races no longer occur over significant portions of areas occupied in earlier times or exist in very low numbers over portions of their present ranges. Thus Townsend's big-eared bat exists in this region only in small isolated pockets of appropriate habitat, and the number of individuals and extent of distribution appear to be shrinking. Many other species are undergoing similar reductions in extent of distribution and in number, and mammalogists are concerned about these trends.

A species is considered endangered when the number of individuals declines to such a low figure that we fear extirpation from a critical portion of the range, or when total extinction is feared. "Endangered" is an official term which provides the species with protection through special laws. Species may be federally classified as Endangered only through action of the U.S. Fish and Wildlife Service. This action means that government projects must not result in further reductions in numbers and that special steps may be taken to protect the species and its habitats. Species not yet in danger of extinction but likely to become endangered are referred to as "Threatened." Federally listed species may be designated Threatened or Endangered throughout their range or only in certain portions of their range. Some states have also generated lists of species considered to be Threatened or Endangered within the boundaries of that state. These species may be provided special protection within the state, or the designation may serve only to alert people to the precarious status of the designated species. Federally Endangered status is noted as appropriate in the species accounts that follow. These designations are not intended to be permanent, for it is hoped that the protection and assistance afforded these species will result in an increase in their numbers sufficient to remove the threat of extinction. To date, however, no mammal species has been removed from the list.

It is clear from this discussion of population trends and of endangered mammals that the preservation of adequate units of undisturbed natural habitats and an unpolluted environment are the major keys to the continued abundance of native mammals. Although major changes are occurring across the landscapes of this region, there is reason for some optimism. The Shenandoah and Great Smoky Mountains national parks in the western portion of the region and the Assateague, Cape Hatteras, and Cape Lookout national seashores along the coast maintain large segments of natural habitats. Numerous national wildlife refuges, wilderness areas, and state parks also protect habitats for many mammals. National and state forests, although managed for timber production, still provide large units of forest habitat and are critical to the survival of many kinds of mammals.

Entrance to a national wildlife refuge, almost always a good place to see mammals.

These public lands, as well as the many private land holdings that are maintained in natural or semi-natural conditions, suggest that the outlook is generally good for many mammalian species as far as habitat is concerned.

The primary danger to all but those species requiring remote undisturbed habitats appears to be related to the spread of environmental pollutants. We must reduce the impact of these pollutants if future generations are to continue to enjoy the varied mammalian fauna present today. Adequate amounts of undisturbed, unspoiled, and unpolluted habitat is the key, as the mammals themselves can solve most other problems. The most restrictive laws governing hunting and trapping will not help if habitats continue to be destroyed and degraded.

Mammals of the Region

Introduction

This section contains accounts for all species of wild mammals known to occur in Maryland, Virginia, and the Carolinas or in the waters of the Atlantic Ocean adjacent to these states. All native terrestrial mammals presently occurring or extirpated and those introduced species that have become established in the region are treated in separate accounts. Related species of marine mammals usually are treated together in a single account.

We have dealt with the mammals of the region at the species level even though many are separable, at least by experts, into two or more races, or subspecies. In some instances, individuals of the same species from mountain and coastal plain populations differ consistently in size and color, and therefore are assigned to different subspecies. These populations, however, may intergrade along a relatively narrow zone of contact where animals of intermediate size and color occur. These differences are noted in the appropriate species accounts. Other differences between subspecies are usually minor, however, and are beyond the scope of this book. Names and distributions of subspecies are shown in Hall's *Mammals of North America* (1981).

A checklist of all mammals treated in the accounts is provided. Common and scientific names are given for each species, both in the checklist and in the species accounts. Common names may be meaningful locally, but

they can lead to confusion because they may vary considerably from one locality to another; for example, the terms "mountain lion," "panther," "painter," "cougar," are used for the same animal. Latinized scientific names, on the other hand, have universal application and are always used in the scientific literature. *Felis concolor* thus identifies the mountain lion, cougar, and so forth, to any scientist in any language. Our use of common names, scientific names, and the systematic sequence of orders, families, and species follows that given in the *Revised Checklist of North American Mammals North of Mexico, 1982* by Jones et al. (1982).

Species accounts include sections entitled "description," "distribution and abundance," "habitat," and "natural history." The paragraphs on description contain brief statements designed to assist the reader in recognizing mammals. These include information on size, common external features, and coloration, but more detailed diagnostic characteristics such as dental formulae and skull measurements are not covered. Such technical data may be found in several of the references listed at the end of this volume.

Pertinent external measurements are provided to give the reader a general indication of the size of each species and to allow size comparisons between or among similar species. Measurements are expressed both in English and metric units, and usually represent minimum and maximum values for mammals in this region. These measurements were derived primarily from *South Carolina Mammals* by Golley (1966) and *Mammals of Maryland* by Paradiso (1969), and thus represent samples of mammals at the southern and northern limits of the four-state region covered in this book. If adequate measurements were not available in these volumes, they were taken from other scientific publications providing measurements of mammals from populations in the mid-Atlantic states or eastern North America. You may, however, encounter individuals smaller or larger than the measurements provided here.

Descriptions of most terrestrial species are accompanied by photographs. Most of the photographs are of individuals from this region, and many were taken with the animals in their natural habitats. Some pictures are of captive animals in zoos or other enclosed habitats created to simulate those in nature as nearly as possible.

Statements on distribution and abundance summarize current information on where each species is found in the region and its level of abundance. A map showing the distribution within the region is provided for most species; of course, within its range animals of a species occur only within suitable habitat, not necessarily throughout the entire area shaded on

the map. Isolated shaded dots represent places from which specimens have been collected outside their normal range. One should remember that both the distribution and abundance of species are dynamic factors that often change with time; for example, the range of the coyote is expanding but that of the fox squirrel is shrinking, and numbers of rabbits and foxes in a population increase and decrease in cycles. Animals caught or observed well beyond their indicated range should be reported to the nearest museum of natural history.

Mammals are usually adapted to specific kinds of habitats, although some, such as the Virginia opossum, are generalists and occupy several different habitat types. Specialists, like some species of cave-dwelling bats, occur only in very specific habitats. The brief habitat descriptions indicate the places in the region where one is most likely to find each species.

Other aspects of the biology of each species are included in the section on natural history. Included in this section is an indication of the foods commonly eaten and brief outlines of daily and seasonal activity cycles, types of "homes" utilized, and reproductive biology. Other remarks relate to the status of each species in natural ecosystems in the region, the major problems faced by the species, and an indication of its interactions with man.

List of Mammal Species Occurring in the Carolinas, Virginia, and Maryland

Following is a checklist of the species of mammals occurring in this region and an indication of their relationships. They are listed by order and family within the Class Mammalia. The common and scientific names are given for each species; introduced, endangered, and extirpated species are noted.

Class Mammalia

ORDER MARSUPIALIA

Family Didelphidae
Virginia opossum—*Didelphis virginiana*

ORDER INSECTIVORA

Family Soricidae
Cinereus or masked shrew—*Sorex cinereus*
Southeastern shrew—*Sorex longirostris*
Water shrew—*Sorex palustris*
Smoky shrew—*Sorex fumeus*
Long-tailed or rock shrew—*Sorex dispar*
Pygmy shrew—*Sorex hoyi*
Northern short-tailed shrew—*Blarina brevicauda*
Southern short-tailed shrew—*Blarina carolinensis*
Least shrew—*Cryptotis parva*

Family Talpidae
Hairy-tailed mole—*Parascalops breweri*

Eastern mole—*Scalopus aquaticus*
Star-nosed mole—*Condylura cristata*

ORDER CHIROPTERA

Family Vespertilionidae
Little brown myotis—*Myotis lucifugus*
Southeastern myotis—*Myotis austroriparius*
Gray myotis—*Myotis grisescens* (Endangered)
Keen's myotis—*Myotis keenii*
Indiana or social myotis—*Myotis sodalis* (Endangered)
Small-footed myotis—*Myotis leibii*
Silver-haired bat—*Lasionycteris noctivagans*
Eastern pipistrelle—*Pipistrellus subflavus*
Big brown bat—*Eptesicus fuscus*
Red bat—*Lasiurus borealis*
Seminole bat—*Lasiurus seminolus*
Hoary bat—*Lasiurus cinereus*
Northern yellow bat—*Lasiurus intermedius*
Evening bat—*Nycticeius humeralis*
Townsend's big-eared bat—*Plecotus townsendii* (Endangered)[a]
Rafinesque's big-eared bat—*Plecotus rafinesquii*

Family Molossidae
Brazilian free-tailed bat—*Tadarida brasiliensis*

[a]Central and eastern races only

ORDER EDENTATA

Family Dasypodidae
Nine-banded armadillo—*Dasypus novemcinctus* (Introduced)

ORDER LAGOMORPHA

Family Leporidae
Marsh rabbit—*Sylvilagus palustris*
Eastern cottontail—*Sylvilagus floridanus*
New England cottontail—*Sylvilagus transitionalis*
Swamp rabbit—*Sylvilagus aquaticus*
Snowshoe hare—*Lepus americanus*
Black-tailed jack rabbit—*Lepus californicus* (Introduced)

ORDER RODENTIA

Family Sciuridae
Eastern chipmunk—*Tamias striatus*
Woodchuck—*Marmota monax*
Gray squirrel—*Sciurus carolinensis*
Fox squirrel—*Sciurus niger* (Endangered)[b]
Red squirrel—*Tamiasciurus hudsonicus*
Southern flying squirrel—*Glaucomys volans*
Northern flying squirrel—*Glaucomys sabrinus*

Family Castoridae
Beaver—*Castor canadensis* (Extirpated and reintroduced)

Family Cricetidae
Marsh rice rat—*Oryzomys palustris*
Eastern harvest mouse—*Reithrodontomys humulis*
Oldfield mouse—*Peromyscus polionotus*

Deer mouse—*Peromyscus maniculatus*
White-footed mouse—*Peromyscus leucopus*
Cotton mouse—*Peromyscus gossypinus*
Golden mouse—*Ochrotomys nuttalli*
Hispid cotton rat—*Sigmodon hispidus*
Eastern woodrat—*Neotoma floridana*
Southern red-backed vole—*Clethrionomys gapperi*
Meadow vole—*Microtus pennsylvanicus*
Rock vole—*Microtus chrotorrhinus*
Woodland vole—*Microtus pinetorum*
Muskrat—*Ondatra zibethicus*
Southern bog lemming—*Synaptomys cooperi*

Family Muridae
Black rat—*Rattus rattus* (Introduced)
Norway rat—*Rattus norvegicus* (Introduced)
House mouse—*Mus musculus* (Introduced)

Family Zapodidae
Meadow jumping mouse—*Zapus hudsonius*
Woodland jumping mouse—*Napaeozapus insignis*

Family Erethizontidae
Porcupine—*Erethizon dorsatum* (Extirpated)

Family Myocastoridae
Nutria—*Myocastor coypus* (Introduced)

ORDER CARNIVORA

Family Canidae
Coyote—*Canis latrans*

[b]Delmarva race only

Red wolf—*Canis rufus* (Extirpated)
Gray wolf—*Canis lupus* (Extirpated)
Red fox—*Vulpes vulpes*
Gray fox—*Urocyon cinereoargenteus*

Family Ursidae
Black bear—*Ursus americanus*

Family Procyonidae
Raccoon—*Procyon lotor*

Family Mustelidae
Fisher—*Martes pennanti* (Extirpated
and reintroduced)
Least weasel—*Mustela nivalis*
Long-tailed weasel—*Mustela frenata*
Mink—*Mustela vison*
Eastern spotted skunk—*Spilogale
putorius*
Striped skunk—*Mephitis mephitis*
River otter—*Lutra canadensis*

Family Phocidae
Harbor seal—*Phoca vitulina*
Harp seal—*Phoca groenlandica*
Hooded seal—*Cystophora cristata*

Family Felidae
Mountain lion—*Felis concolor*
(Endangered)[c]
Bobcat—*Felis rufus*

ORDER MYSTICETI

Family Balaenopteridae
Minke whale—*Balaenoptera
acutorostrata*
Sei whale—*Balaenoptera borealis*
(Endangered)
Bryde's whale—*Balaenoptera edeni*
Fin whale—*Balaenoptera physalus*
(Endangered)
Blue whale—*Balaenoptera musculus*
(Endangered)

Humpback whale—*Megaptera
novaeangliae* (Endangered)

Family Balaenidae
Black right whale—*Balaena glacialis*
(Endangered)

ORDER ODONTOCETI

Family Delphinidae
Rough-toothed dolphin—*Steno
bredanensis*
Bottle-nosed dolphin—*Tursiops
truncatus*
Bridled spotted dolphin—*Stenella
frontalis*
Atlantic spotted dolphin—*Stenella
plagiodon*
Striped dolphin—*Stenella
coerulcoalba*
Long-snouted spinner dolphin—
Stenella longirostris
Saddle-backed dolphin—*Delphinus
delphis*
Atlantic white-sided dolphin—
Lagenorhynchus acutus
Risso's dolphin or grampus—
Grampus griseus
False killer whale—*Pseudorca
crassidens*
Long-finned pilot whale—
Globicephala melaena
Short-finned pilot whale—
Globicephala macrorhynchus
Melon-headed whale—
Peponocephala electra
Killer whale—*Orcinus orca*

Family Phocoenidae
Harbor porpoise—*Phocoena
phocoena*

Family Ziphiidae
Goose-beaked whale—*Ziphius
cavirostris*

[c]Eastern races only

True's beaked whale—*Mesoplodon mirus*
Gervais' beaked whale—*Mesoplodon europaeus*
Dense-beaked whale—*Mesoplodon densirostris*

Family Physeteridae
Dwarf sperm whale—*Kogia simus*
Pygmy sperm whale—*Kogia breviceps*
Sperm whale—*Physeter macrocephalus* (Endangered)

ORDER SIRENIA

Family Trichechidae
Manatee—*Trichechus manatus* (Endangered)

ORDER ARTIODACTYLA

Family Suidae
Wild pig—*Sus scrofa* (Introduced)

Family Cervidae
Wapiti or elk—*Cervus elaphus* (Extirpated)
Sika deer—*Cervus nippon* (Introduced)
White-tailed deer—*Odocoileus virginianus*

Family Bovidae
Bison—*Bison bison* (Extirpated)

Pouched Mammals
Order Marsupialia

Most species of marsupials, including the familiar kangaroo, koala, bandicoot, and wombat, inhabit Australia, Tasmania, New Guinea, and nearby islands. A less diverse marsupial fauna occurs in North and South America. The only marsupial in the United States is the Virginia opossum, in the Family Didelphidae.

Marsupials are most distinct from other live-bearing mammals in that pregnant females have a less specialized placenta, which in more advanced orders provides support and nourishment for embryos that are born more fully developed. The primitive marsupial placenta provides little support, and embryos are retained only briefly in the uterus of the mother; they are born incompletely developed, or in a "larval" state. The females of most marsupials possess an abdominal pouch, or marsupium, into which the newborn move, attach to a teat to suckle milk, and complete their development.

Virginia Opossum
Didelphis virginiana

Description. The Virginia opossum is a robust, heavy-bodied animal, about the size of a domestic cat but with shorter legs. Adults from the four-state region average about 28 inches (72 cm) in total length, including a tail of about 12 inches (30 cm). Males weigh about 5 pounds (2.3 kg) and females 4 pounds (1.8 kg). The underfur is dense, with hairs that are white basally and black at the ends, interspersed with long white guard hairs—hence, the animal's gray, grizzled appearance. The face and toes are white. The head is conical with a slender, pointed snout. The ears are naked, leathery, and black in color. The tail is scantily haired, pale in coloration, and prehensile. The forefeet have claws on all 5 toes, but the hind feet have an enlarged first toe that is thumb-like, opposable, and lacks a claw. Females have a fur-lined abdominal pouch; the usual number of teats in the pouch is 13, but may vary from 9 to 17.

Distribution and Abundance. The Virginia opossum is abundant in the southeastern United States and is found throughout the Carolinas, Virginia, and Maryland.

Habitat. The species is present in a wide variety of habitats from relatively dry upland areas to those of considerable wetness, but it prefers wooded bottomlands near streams, ponds, swamps, and other sources of water. It is more likely to be found where woodland is interspersed with meadows or fields rather than in extensive, dense forests. It is adaptable to human presence and is not uncommon around both rural and town residences.

Virginia opossum (Didelphis virginiana).

Natural History. Home for a Virginia opossum is almost any dry, sheltered place, such as a hollow tree, fallen log, cavity in a rock or brush pile, or the den or nest of another animal. Home sites are not permanent and may be changed frequently. The animal makes a nest of leaves, grass, or similar material, which it collects with its mouth and front feet, passes beneath the body to the hind feet and tail, and then carries to the nest as a bundle in a loop formed by the end of the tail.

Opossums are good but somewhat clumsy climbers, using the opposable toe on the hind foot and the prehensile tail in climbing. They are most often seen searching for food on the ground, where they have a slow, ambling gait. They also enter water and are strong but slow swimmers.

Mating in Virginia opossums extends from late January to early July.

Two litters are produced each year, though more than two may be produced if earlier litters are lost. Development in the uterus is brief, and young are born 12 to 13 days after copulation; they are weaned after about 100 days. Thus, a female that mates and produces a litter in Febru-

ary typically will mate again in May or early June. The mother may give birth to as many as 18 to 21 larval offspring; however, some fail to reach her pouch or complete their development, resulting in an average litter size of 7 to 9 individuals.

At birth, the tiny young are blind and naked; the forelimbs, mouth, and sense of smell are well developed, but the hind limbs and other structures remain rudimentary. Each uses its front feet and claws to crawl 2 to 3 inches (5 to 8 cm) from the mother's genital opening to the pouch, and attaches to a nipple. The mother provides no assistance except for a path of fur dampened by her tongue. They remain attached for about 60 days and after 80 days leave the pouch briefly. As they become larger, they alternate between riding in the pouch and clinging to the fur on the mother's back. Virginia opossums are capable of mating the year after birth.

Virginia opossums are shy, secretive, and unaggressive animals and usually attempt to escape when mo-

lested. However, defensive behavior is highly developed, consisting of displays intended to bluff and intimidate. If all else fails, the animal may feign death, or "play possum"—it curls on its side, becomes limp, closes its eyes, lolls its tongue from its mouth, and the heartbeat slows appreciably. This reaction may be brief or last up to 6 hours.

These animals are most active in spring and summer; activity declines markedly in cold weather. They tend to be nomadic and to lead solitary lives, except for mating. They are almost exclusively nocturnal, with movements concentrated around a den or between den sites. The Virginia opossum prefers to eat animal matter such as insects, earthworms, land snails, amphibians, and carrion. It also eats plant material, such as fruits and grains, especially in fall and winter. It has numerous predators, including a variety of carnivorous mammals and birds, but automobiles may be its worst enemy. Its meat is valued by some people, as is its fur.

Shrews and Moles
Order Insectivora

The insectivores, in contrast to the marsupials, are true placental mammals, as are all other mammals occurring in the Carolinas, Virginia, and Maryland. These mammals develop their young entirely within the body of the female and have a well-developed placenta; there is no abdominal pouch as in opossums.

Insectivores are found in all parts of the world except Australia, Greenland, most of South America, and the polar regions. The order consists of 7 families, only 2 of which occur in North America: Family Soricidae, which in the Carolinas, Virginia, and Maryland includes 9 species of shrews, in 3 genera; and Family Talpidae with 3 species of moles, in 3 genera, in the region. No single character or combination of characters affords clear distinction between all insectivores and all other mammals. Of modern insectivores, shrews are the most generalized in structure and habits.

Shrews and moles are small animals, the smallest being the pygmy shrew weighing about ⅛ ounce (2.5 gm). The snout is elongate, flexible, and pointed, and highly sensitive. The ears and eyes are small, almost rudimentary. The teeth are simple, there being little differentiation between canines and cheek teeth; all are sharp and pointed. The incisor teeth tend to extend forward, an aid in capturing insects and other prey. The skull is wedge-shaped, and cheek bones are either lacking, as in shrews, or much reduced, as in moles. Each foot has 5 digits with claws; feet are plantigrade, or entirely on the ground when walking. Most species have prominent scent glands on the flanks, producing a strong, musky odor.

Moles are fossorial, tunneling extensively through the soil and spending most of their time there; shrews, in contrast, primarily live at the surface under cover of grasses, leaves, logs, and rocks. Structural adaptations of moles for burrowing include greatly enlarged forefeet and shoulder girdles, cylindrical bodies, reduced hind feet and pelvic girdles, and short, thick fur which accommodates either forward or backward movement in the tunnel. The smaller shrews have all 4 feet of equal size, are less densely furred, and are more generalized in structure.

Most insectivores are ferocious predators and have almost insatiable appetites; they appear to search for food almost constantly. They experience a high rate of heat or energy loss from the body due to the relatively greater surface area compared to body mass or weight. A mole or shrew deprived of food for only a few hours will usually die.

Shrews (with 6 species in the genus *Sorex*, 2 in *Blarina*, and 1 in *Cryptotis*) are often difficult to identify due in part to their small size and similar appearance. An effort has been made in the accounts that follow to provide descriptive statements that distinguish the various species; however, it may be necessary to seek help from someone familiar with this group of mammals. In contrast to shrews, the 3 species of moles in the region present little problem in their identification.

Cinereus or Masked Shrew

Sorex cinereus

Description. Shrews of the genus *Sorex* can be distinguished from short-tailed (*Blarina*) and least (*Cryptotis*) shrews by their relatively long tails, which are more than half the head and body length. The masked shrew has minute eyes, a long pointed nose, and small ears that are nearly concealed by fur. The pelage in winter is dark grayish brown above and light grayish brown

below, whereas in the summer it is more brownish; the tail is distinctly bicolored (dark above and pale beneath). The masked shrew most closely resembles the southeastern shrew, but the masked shrew is somewhat larger in overall size, has a longer and hairier tail, and is slightly darker in pelage coloration. Its total length ranges from 3½ to 4⅛ inches (88 to 104 mm), including a tail of 1¼ to 1⅝ inches (31 to 40 mm); the length of the hind foot ranges from ⅜ to ½ inch (10 to 13 mm).

Distribution and Abundance. The masked shrew is found throughout most of Maryland and the mountains of Virginia and the Carolinas. This secretive little mammal is abundant where it occurs, but the number of individuals in a local population varies greatly. Some years many animals oc-

Cinereus or masked shrew (Sorex cinereus).

cur in habitats where at other times they are quite rare.

Two races of masked shrew occur in the region, a larger form in the western mountains and a smaller variety with a shorter tail in north-central Maryland and the Eastern Shore. These races have been regarded as distinct species by some biologists.

Habitat. No other species of *Sorex* occurs in as many different habitats as the masked shrew. It is most abundant in moist deciduous, coniferous, or mixed forests where moss-covered rocks, decaying logs and stumps, and leaf litter provide suitable ground cover. Other habitats in which the masked shrew has been taken include grassy fields, sphagnum bogs, swamps, and meadows.

Natural History. The masked shrew reproduces from March to September, and several litters are born during that time. Four to 10 embryos have been counted, but 5 or 6 young per litter are probably most common. The gestation period lasts 22 days, the young nurse for 21 days, and reproductive maturity is attained in 5 to 6 months. They seldom live 2 years in nature.

Globular nests are built under logs, stumps, roots, or rocks; they are approximately 3 inches (76 mm) in diameter and built of dried leaves and grasses. The masked shrew uses the surface and subterranean runways of several species of rodents, and it also constructs its own. It moves around these runways in search of prey such as adult and larval insects, earthworms, other shrews and small mice, and snails and slugs; some vegetable matter is eaten also. Owls, foxes, and weasels are their primary predators.

This shrew is of great economic importance to humans because it consumes large numbers of arthropods. These tiny bundles of energy are active throughout the day and night and eat more than their weight in food during a 24 hour period.

Southeastern Shrew
Sorex longirostris

Description. Like other members of the genus *Sorex*, the southeastern shrew has a relatively long tail, small eyes, an elongate rostrum, and minute ears that are mostly hidden by fur. This medium-sized *Sorex* is similar to the masked shrew in body dimensions; however, the southeastern shrew is slightly smaller than the masked shrew in most external measurements, particularly the length of tail and hind foot, and its tail is not noticeably bicolored. The fur of the southeastern

Southeastern shrew (Sorex longirostris).
Photograph by Thomas W. French.

shrew is reddish brown on the back
and slightly paler on the belly, whereas
that of the masked shrew is grayish
brown dorsally. Other species of *Sorex*
are distinctly smaller (the pygmy
shrew) or larger (long-tailed, water,
and smoky shrews) than the southeastern shrew in body size. The total
length of this shrew ranges from 2⅞
to 4⅝ inches (72 to 116 mm), including a tail of ⅞ to 2 inches (21 to 51
mm); the length of the hind foot
ranges from ⅜ to ½ inch (10 to 14
mm).

Distribution and Abundance. The
southeastern shrew is distributed
from the Western Shore of Maryland
southward throughout most of Virginia and the Carolinas, including
some of the larger islands along the
Outer Banks. It is the only *Sorex*
known to occur at lower elevations in
the Carolinas, and its range appar-

ently seldom overlaps that of the
masked shrew. Once considered rare,
our investigations indicate that the
southeastern shrew is locally abundant
in suitable habitat.

The southeastern shrew also inhabits the Dismal Swamp region of
southeastern Virginia and northeastern North Carolina. Here it is
larger and somewhat duller and more
brownish in color than elsewhere in
its geographic range; consequently, it
might be confused with the masked
shrew. The southeastern shrew, however, is the only *Sorex* in the swamp
area.

Habitat. Damp fields, canebrakes,
thickets, and lowland forests are the
preferred habitats of southeastern
shrews, particularly under tangles of
honeysuckle, poison ivy, and other
vines. Relatively dry upland fields and
forests also are inhabited.

Natural History. Until recently few
specimens of the southeastern shrew
had been collected, so several aspects

of their natural history are poorly known. Biologists are now catching them in cans buried flush with the ground, and as a result our knowledge is increasing.

Southeastern shrews reproduce from March through October; it is probable that 2 litters are born during that time, each consisting of 1 to 6 offspring. Females may become reproductively active in their first year. Their nests consist of shallow depressions lined with dried leaves and grasses, and usually are located under or within rotting logs.

These shrews consume many invertebrates such as spiders, caterpillars, crickets, beetles, centipedes, slugs, and snails; some vegetation also is eaten. They are prey for owls, snakes, opossums, and domestic dogs and cats.

Water Shrew
Sorex palustris

Description. The dense fringe of stiff hairs along the side and toes of each hind foot readily distinguishes the water shrew from other species of shrews in the region; further, the hind feet are enlarged, measuring about ¾ inch (20 mm) in length, and have webbing between the third and fourth toes. The water shrew is the largest *Sorex* present in the region, having a total length of approximately 6 inches (153 mm). It is nearly as large as the northern short-tailed shrew, but these species can be separated by the longer tail of the water shrew, which measures about 2⅝ inches (68 mm) or more than two-thirds the head and body length. The pelage of the water shrew is blackish above and slightly paler on the underside; the tail is bi-

Water shrew (Sorex palustris). *Photograph by Roger W. Barbour.*

colored, also dark above and pale below.

Distribution and Abundance. The water shrew is known to occur in this region in only 2 isolated localities in the mountains of western North Carolina (Clay County) and Virginia (Bath County), but it has also been taken near Cranesville, West Virginia, on the Maryland border. It is listed as an Endangered Species in Virginia, and a similar listing in North Carolina would be prudent. Efforts are needed to locate additional populations in the southern Appalachians, and all should be protected.

Habitat. Banks of flowing streams and creeks in forests of sugar maple, beech, yellow birch, and rhododendron, or bogs with spruce, hemlock, and willow are ideal for water shrews in this area.

Natural History. As its common name implies, the water shrew is well adapted to a semiaquatic existence. It uses its fringed and webbed hind feet as paddles in swimming and has even

been observed using them to run across the water's surface. While swimming under water, this shrew uses its nose to search for food between and under gravel and rocks; it returns quickly to the surface by means of air trapped in the fur and under the hind feet. Its diet consists primarily of aquatic animals such as larval and adult insects, leeches, snails, small fish, and fish eggs. The reproductive season of the water shrew extends from March until August. Four to 8 young are born after a gestation period of about 21 days.

Smoky Shrew
Sorex fumeus

Description. The smoky shrew and the long-tailed shrew are similar in being the only shrews in the region that are medium-sized and possess relatively long tails; specifically, the smoky shrew measures from 4⅛ to 4¾ inches (104 to 120 mm) in total length, including a tail of 1⅝ to 2 inches (40 to 51 mm). In contrast, least and short-tailed shrews have much shorter tails, and other species of *Sorex* are either larger (the water shrew) or smaller (masked, southeastern, and pygmy shrews) in size.

Smoky and long-tailed shrews differ in that the former has a scantily-haired tail that is about two-thirds the head and body length and narrow in diameter, whereas that of the latter is more than four-fifths the head and body length, moderately haired, and thick. Also, the smoky shrew has a greater average weight than the long-

Smoky shrew (Sorex fumeus).

tailed shrew, and its tail is darker above and paler beneath rather than dark all over. The dorsal fur of the smoky shrew is reddish brown during the summer and grayish brown in the winter, but always darker than the grayish belly. The hind foot measures about ½ inch (13 to 14 mm) in length.

Distribution and Abundance. The Appalachian Mountains in western Maryland, Virginia, and the Carolinas are the home of the smoky shrew. In suitable habitat it is locally abundant, but population numbers vary from season to season and year to year.

Habitat. The smoky shrew is a denizen of cool, moist mountain forests where it occupies the forest floor under a thick ground cover of leaf litter and moss-covered rocks, logs, and stumps. Bogs and roadside cuts that

expose bare rock faces are inhabited by this relatively widespread species as well, but birch and hemlock forests seem to be preferred.

Natural History. Small spherical nests of shredded leaves and grasses are built under or within fallen logs and

stumps, in rock crevices, or under rocks. Here as many as 3 litters are born between March and early October after a gestation period of about 20 days. Litters usually contain 5 or 6 young, but from 2 to 10 have been noted. Sexual maturity is attained after the first winter.

Smoky shrews eat salamanders and insects, centipedes, earthworms, and other small invertebrates that inhabit the forest floor. They traverse the runways of moles, voles, and short-tailed shrews in search of food, and act as a natural biological control of insects. They are therefore of great economic importance to humans. Owls, hawks, bobcats, weasels, and foxes prey upon them, and they apparently do not live much more than 2 years in nature.

Long-tailed or Rock Shrew
Sorex dispar

Description. The long-tailed shrew, a *Sorex* of medium size and therefore similar to the smoky shrew, is separated easily from the much larger water shrew and distinctly smaller masked, southeastern, and pygmy shrews. Compared to the smoky shrew, the long-tailed shrew has a much longer and thicker tail, being more than four-fifths (rather than about two-thirds) the head and body length. The dorsal fur of the long-tailed shrew is slate gray throughout the year, and that of the belly is only slightly paler. The dark tail is moderately haired and not noticeably bicolored. The total length of the long-tailed shrew is from 4½ to 5¼ inches (114 to 132 mm), including a tail of 1⅞ to 2⅜ inches (49 to 59 mm). The length of the hind foot is about ½ to ⅝ inch (14 to 15 mm) long. Specimens from North Carolina are slightly larger and darker than those from Virginia and Maryland.

Distribution and Abundance. Specimens of the long-tailed shrew have been taken from the Appalachian Mountains of western North Carolina, Virginia, and Maryland. This shrew is extremely habitat specific and is, therefore, not widely distributed in the region. It is common in suitable habitat, especially in Virginia and farther north, but in North Carolina it is rare and localized.

Habitat. The deepest crevices in cliffs, rocky slopes, and talus in cool, moist forests are the favorite haunts of the long-tailed shrew, but moss-covered rocks and logs adjacent to mountain streams also provide suitable habitat.

Natural History. A scant amount of information about the biology of the

long-tailed shrew has been gathered.
Individuals hunt in rock crevices for
centipedes, insects, and spiders.
Three to 5 young are born in the
months from May to August. The
long-tailed shrew, like all shrews in
the region, is beneficial to humans because it consumes many invertebrates,
some of them harmful.

Pygmy Shrew
Sorex hoyi

Description. This is the smallest mammal in North America, weighing about
as much as a dime. The pygmy shrew
has a relatively long tail (slightly more
than one-half the head and body
length), a long pointed snout, minute
eyes and ears, and small feet. The
masked shrew and southeastern shrew
are the only *Sorex* that approach the
pygmy shrew in size, but they are
larger. In color the pygmy shrew is
dark reddish brown above in summer
pelage and grayish brown above in the
winter; the belly is much paler. Its total length is from 2¾ to 3⅜ inches
(70 to 86 mm), including a tail of 1 to
1¼ inches (25 to 33 mm); the length
of the hind foot ranges from ¼ to ⅜
inch (7 to 10 mm).

Distribution and Abundance. Fewer
than 2 dozen specimens of the pygmy
shrew are known from the hardwood
forests in Maryland, Virginia, and
North Carolina; it is one of the rarest
mammals in the region. Documenta-
tion of its distribution in the region is
far from complete, and to date specimens have not been taken in South

Carolina, the piedmont and coastal
plain of North Carolina, the southern
tier of counties in central and eastern
Virginia, or the Eastern Shore of Virginia and Maryland.

Habitat. Pygmy shrews have been
taken along ridges and slopes in deciduous forests with scattered rocks,
fallen logs, and leaf litter covering the
forest floor. Most specimens have
been captured inside fallen logs or in
cans buried flush at ground level.

Natural History. Due to its scarcity in
the region, little is known about the
pygmy shrew. It apparently is active
both day and night in search of grubs,
earthworms, and insects. Pygmy
shrews molt in the spring and reproduce in the summer. A litter may have
from 5 to 8 young. A well-developed
gland on the flank secretes a odorifer-
ous substance probably used in marking territory and attracting mates.

Although the pygmy shrew is undoubtedly rare, it is probably more
common than our meager records indicate. All small shrews should be

carefully identified in hopes of providing additional and valuable information about this relatively rare mammal.

Northern Short-tailed Shrew
Blarina brevicauda

Description. Shrews of the genus *Blarina* are relatively large with minute eyes and a short tail; the ears are reduced in size and hidden by short, dense, velvety fur. The tail of short-tailed shrews is approximately one-third the head and body length, whereas that of long-tailed shrews (genus *Sorex*) is greater than one-half the head and body length. Short-tailed shrews are most similar to least shrews in external morphology, but the former are larger than the latter in most dimensions. The northern short-tailed shrew is 3¾ to 5⅜ inches (95 to 136 mm) in total length, including a tail of ¾ to 1⅛ inches (19 to 30 mm), and a hind foot of ½ to ⅝ inch (13 to 17 mm).

It is difficult to distinguish between the northern and southern short-tailed shrews, and experts currently are studying both. The northern short-tailed shrew averages larger than its southern relative, but some individuals of both species overlap in size and can easily be misidentified. In this region, a combination of factors best serves to separate them. First, the limits of their distribution do not overlap except in the tidewater regions of eastern North Carolina and Virginia; therefore, specimens can be identified by location of capture except when from this area. Second, in the area where the distributions of

Northern short-tailed shrew (Blarina brevicauda).

both species overlap, the northern short-tailed shrew averages larger in size than the southern short-tailed shrew and is usually darker in pelage coloration. Some specimens, however, are so similar that they can be identified only by an expert.

Northern short-tailed shrews from the Dismal Swamp in Virginia and the coastal plain of North Carolina are intermediate in size and grayish brown in color, those from the Carolina mountains are large and slate black, and those from Maryland and the remainder of Virginia are smaller in size and intermediate in color.

Distribution and Abundance. The northern short-tailed shrew is one of the most abundant small mammals in the region. It occurs throughout Maryland and most of Virginia; in North Carolina it is known from the eastern coastal plain and western mountains, but not the central piedmont, and in South Carolina it occurs only in a few of the westernmost counties.

Habitat. Most terrestrial environments, including salt marshes, fields, and forests, are inhabited by the northern short-tailed shrew, especially those areas with a thick layer of leaf litter. Short-tailed shrews prefer to tunnel under the litter; consequently, it is seldom encountered by humans except when captured by the family cat.

Natural History. At various times in the past, the northern and southern short-tailed shrews have been considered to represent a single species; however, biologists recently concluded that 2 distinct species exist. Many aspects of their natural history are similar, and those topics are discussed in the account of the southern short-tailed shrew.

Southern Short-tailed Shrew
Blarina carolinensis

Description. The southern short-tailed shrew closely resembles its northern relative, the northern short-tailed shrew, and the least shrew in body proportions. Tail length in these 3 species is about half the length of the head and body, the eyes are minute, and the ears are hidden by the short, dense fur. Means by which the southern short-tailed shrew can be separated from the northern short-tailed shrew and the least shrew are discussed in the accounts of those species. The southern short-tailed shrew measures 3⅛ to 4⅞ inches (79 to 123 mm) in total length, including a tail of

Southern short-tailed shrew (Blarina carolinensis).

½ to 1⅛ inches (14 to 29 mm); the hind foot is ⅜ to ⅝ inch (8 to 15 mm) long.

Distribution and Abundance. This shrew is distributed from south-central and portions of eastern Virginia southward throughout the piedmont and coastal plain of the Carolinas. Although it is relatively common in this area, it travels beneath the leaf litter and is observed infrequently.

Habitat. Because leaf litter affords protection to this secretive little mammal, habitats such as forests are preferred. Individuals, however, have been observed in grassy fields, bogs, meadows, and tidal marshes.

Natural History. Southern and northern short-tailed shrews are similar in several aspects of their natural history. Both move either in surface runways or under dead leaves and grasses on the ground. Runways used by short-tailed shrews are frequently those used by various species of mice and voles, and the shrews contribute little in runway maintenance. Nests of dried grasses and shredded leaves,

each 4 to 6 inches (10 to 15 cm) in diameter, are built under fallen logs and stumps.

Reproductive patterns are similar in both southern and northern short-tailed shrews. Females produce 3 or 4 litters each year during the months from February to November. The gestation period is approximately 3 weeks, and a litter may consist of 1 to 10 young, but 6 or 7 are typical. The young are born hairless and pink. They are weaned at about 3 weeks of age.

A common trait of short-tailed shrews is the unmistakably strong odor produced by a gland on each side near the hip. The odor is particularly evident during the breeding season and is used to communicate the reproductive status and territorial boundaries of individual shrews. A variety of clicks and squeaks also are used in communication, as is echolocation.

Short-tailed shrews secrete a toxin from the salivary glands that paralyzes prey such as earthworms, insects, spiders, centipedes, small vertebrates, snails, and slugs. Paralyzed animals sometimes are stored, but these shrews have such voracious appetites that they eat at least their total weight in food each day. Vegetation is also consumed in small quantities.

Short-tailed shrews have no direct impact on humans. Both species are beneficial because they consume so many arthropods and are, therefore, particularly important in insect control.

Least Shrew
Cryptotis parva

Description. This shrew has a long nose, small eyes, inconspicuous ears, and a short tail. The fur is brownish gray above and paler below, with a silver frosting on each hair; sometimes there is a yellowish patch of fur on the chest in front of the forelegs. The least shrew can be separated easily from members of the genus *Sorex* by the length of its tail, which is much less than half the head and body length. As least shrews resemble a miniature short-tailed shrew in body proportions and pelage coloration, the two species are sometimes hard to separate without the assistance of an expert; however, even the largest least shrew seldom exceeds the smallest short-tailed shrew in size. Least shrews measure from 2¾ to 3⅝ inches (70 to 92 mm) in total length, including a tail of ½ to 1 inch (13 to 26 mm); the hind foot ranges from ⅜ to ½ inch (9 to 13 mm) in length.

Distribution and Abundance. The least shrew is a common mammal found throughout the region, but it is seldom observed because of its secretive habits and diminutive size. Local populations increase in number when living conditions are ideal, but decline when they are severe, resulting in great fluctuations in numbers from year to year and even season to season.

Habitat. A wide variety of habitats are frequented by the least shrew, but relatively open areas dominated by herbaceous vegetation, such as grassy fields and salt marshes, are preferred.

Least shrew (Cryptotis parva).

In the Carolinas it also inhabits the maritime forests of the coastal barrier islands, but numbers there are low.

Natural History. Nests are 2 to 5 inches (5 to 13 cm) in diameter, composed of dried grass and shredded leaves, and usually located under flat

rocks, logs, or stumps, but sometimes under pieces of tin or sheet metal. Least shrews are somewhat gregarious, and more than 2 dozen shrews have been reported to occupy a single nest. This is noteworthy inasmuch as most shrews are not thought to be colonial. Least shrews communicate by a complex assortment of squeaks and clicks.

A single female produces several litters between March and December. Each litter consists of 2 to 7 offspring, the average being 4 or 5. The offspring are born approximately 3 weeks after copulation; they are pink and naked at birth but grow rapidly. Their eyes open at 2 weeks, and they are weaned at 3 weeks of age.

The tiny mammal is beneficial because it consumes many insects. It also frequently consumes other terrestrial arthropods, earthworms, and

snails and occasionally eats small vertebrates. On the other hand, least
shrews are a major source of food for
many species of owls, as well as other
predators such as hawks, snakes,
foxes, skunks, and domestic dogs and
cats.

The least shrew does well in captivity, and several aspects of its biology
are relatively well known. It meticu-
lously cleans itself and keeps its nest
free of excrement, and individuals
have lived almost 2 years in captivity.

Hairy-tailed Mole
Parascalops breweri

Description. As its name implies, the
tail of this mole is covered by long
bristly hairs, a character that distin-

guishes it from other species of moles.
The hairy-tailed mole is 5½ to 6⅝
inches (139 to 169 mm) in total
length, and the relatively short tail of
⅞ to 1⅜ inches (23 to 36 mm) is approximately one-fourth the head and
body length. The forefoot of the
hairy-tailed mole is as broad as it is
long, the toes are not webbed, the
eyes are reduced, and external ears
are absent. The dense fur is soft and
silky, blackish above and somewhat
paler below, but sometimes with small
irregular spots of white; the hair of
the nose, feet, and tail becomes white
with age. Males are slightly larger
than females in most dimensions.

Distribution and Abundance. The hairy-
tailed mole is relatively abundant in
suitable habitat in the mountainous
regions of western Maryland, Virginia,
and North Carolina. It may be more
common than records indicate, but

Hairy-tailed mole (Parascalops breweri).

due to its fossorial nature, it is difficult to trap and usually is encountered by humans only as it moves across roads and forages above ground.

Habitat. Hardwood and coniferous forests are the most frequent haunts of hairy-tailed moles, but pastures, cultivated fields, grassy roadsides, and rhododendron thickets also provide suitable habitat. The extensive tunnel systems of this mole, scarcely obvious above ground, are usually found in well-drained loamy soils. There apparently is an altitudinal separation between the hairy-tailed mole and the eastern mole—the former occurs above 2,000 feet (610 m) in the four-state region, whereas the latter generally occupies lower elevations.

Natural History. Hairy-tailed moles dig tunnels both near the surface and at depths of 10 to 20 inches (25 to 50 cm) below the surface. Deep tunnels are used during the winter, whereas in warmer months surface tunnels are traveled more frequently. Spherical nests, about 6 inches (15 cm) in diameter and composed of dried leaves and grasses, are built in the deeper tunnels.

These moles are solitary during most of the year, but in early spring individuals of both sexes live in the same tunnel system during their mating period. Females produce a single litter each year approximately 1 month after fertilization; a litter usually contains 4 or 5 young. The young are born naked except for short whiskers. They leave the nest after a month and become sexually mature at about 10 months of age.

Earthworms, ants, beetles, and other terrestrial arthropods are important food items, but soil and small roots are consumed incidentally as well. Predators include red foxes, snakes, opossums, and even bull frogs! Hairy-tailed moles are beneficial because they till the soil and eat pest insects; however, they also damage lawns, gardens, and golf courses with their extensive burrow systems.

Eastern Mole
Scalopus aquaticus

Description. The greatly enlarged forefeet, noticeably wider than long, readily identify the eastern mole; also, it lacks the nasal rays of the star-nosed mole, and its tail is essentially naked and not densely furred as in the hairy-tailed mole. The eyes of the eastern mole are reduced in size and covered by a thin layer of skin; there are no external ears, and the toes of the front and hind feet are webbed.

The dense, soft fur is variable in

Eastern mole (Scalopus aquaticus).

color, ranging from silver gray in specimens from the coastal plain of the Carolinas to brownish black in those from Maryland, Virginia, and the southern Appalachian Mountains. The fur on the belly is always paler than that on the back. Some specimens have small patches of white fur; these are especially prevalent on the nose. Others, particularly reproductively active males, have fur that is stained with a yellowish tint from the secretions of various skin glands.

Eastern moles are 5⅝ to 7⅜ inches (144 to 187 mm) in total length, and the tail is ¾ to 1¼ inches (18 to 33 mm) in length. The darker moles of Maryland, Virginia, and western Carolina are slightly larger in size than the paler ones from the piedmont and coastal plain of North and South Carolina. Males average somewhat larger than females.

Distribution and Abundance. The eastern mole is the most widespread and abundant mole in the four-state region, occurring everywhere except the Allegheny Mountains of western Maryland. It is common on many of the larger barrier islands along the

Outer Banks, but becomes rare and localized in the southern Appalachian Mountains where it is restricted to river bottoms and other low-lying habitats.

Habitat. Eastern moles inhabit almost any type of environment, provided the soil is a well-drained loam or sand; soils with large amounts of clay or gravel are avoided. Grassy fields, meadows, pastures, and broken forests harbor the greatest concentrations of moles, and lawns, gardens, and golf courses are damaged occasionally by the extensive tunnel systems.

Natural History. Two types of tunnels are excavated by eastern moles. Those immediately beneath the surface of the soil are used temporarily when searching for food such as earthworms, various kinds of insects, and even plant material. Permanent tunnels are located 6 to 24 inches (15 to 60 cm) deep and provide shelter for nests and passage to feeding areas. Deeper tunnels are utilized throughout the year, whereas surface tunnels are built and maintained during the spring, summer, and fall. Damage to tunnels is repaired by the animals as soon as it is discovered.

Nests of grass and leaves are built 5 to 18 inches (13 to 46 cm) beneath the surface of the soil, usually under a stump or clump of roots. They measure 7 to 9 inches (18 to 22 cm) in length and 4 to 5 inches (10 to 12 cm) in width. Several nests are maintained in the summer, but it is thought that only one is used in the winter.

Individuals remain solitary except during the reproductive season. A single litter is produced annually; it consists of 2 to 5 young which are born in early spring after a gestation period of 4 to 6 weeks. The young are naked at birth, but grow quickly and leave the nest at 4 weeks of age. Females are not capable of breeding until after their first winter.

Star-nosed Mole
Condylura cristata

Description. The star-nosed mole can be readily distinguished from other species of moles by the 22 fleshy appendages that surround the nostrils. The short, dense fur is blackish brown to black on the back and somewhat paler on the belly. In total length, the star-nosed mole measures from 6¼ to 7¼ inches (158 to 185 mm); the tail is moderately haired and 2¼ to 2⅝ inches (57 to 66 mm) in length, or approximately as long as the head and body; and the relatively large hind foot measures between 1 and 1⅛ inches (25 to 28 mm) in length. Specimens become progressively smaller from Maryland to South Carolina, but males and females are similar in size at any one locality. In both sexes the tail thickens in the winter and spring months from an accumulation of fat.

Distribution and Abundance. The star-nosed mole is locally common throughout Maryland and northern Virginia and southward along the Appalachian Mountains; it occurs sporadically and is uncommon to rare in the coastal plain of southern Virginia and the Carolinas. No specimens have

Star-nosed mole (Condylura cristata). *Photograph by Dwight R. Kuhn.*

been taken in the piedmont of southern Virginia or North Carolina.

Habitat. Moist meadows, fields, swamps, and woods are the preferred habitat of star-nosed moles. Burrows that are near bogs and streams sometimes lead directly into water. Star-

nosed moles are excellent swimmers and divers, but they also travel above ground, or over and below snow, especially during the winter when deeper soils may become frozen.

Natural History. Star-nosed moles are active during the day and night, and throughout the year. In contrast to the predominantly subterranean tunnels constructed by other species of moles, the burrows of the star-nosed mole alternate between underground tunnels and short surface runways. Star-nosed moles appear to be gregarious and several moles may use adjoining tunnels.

Nests are roughly spherical in form, composed of dead leaves and grasses, and are 5 to 6 inches (13 to 15 cm) in diameter. They are built just beneath the surface of the ground, usually under a log, stump, or root, but always above the high water level. The star-nosed mole feeds primarily on aquatic and terrestrial worms and insects, crustaceans, and small fish, using its sensitive nasal rays to probe the environment.

Males and females pair in the winter and mate in the spring; a single litter is produced each year. Three to 7 young are born between April and June after a gestation period of approximately 45 days. The young are born hairless, but grow rapidly and leave the nest in 3 to 4 weeks. Males and females become sexually mature at about 10 months and breed in the first year after birth.

The list of animals that prey on star-nosed moles is extensive, including various hawks and owls, house cats, striped skunks, mink, corn snakes, and largemouth bass. Even an eastern chipmunk has been seen eating one! This mole damages low-lying lawns and golf courses occasionally; however, it is beneficial because it consumes many insects and aerates the soil.

Bats
Order Chiroptera

Bats are the only mammals capable of sustained flight. For many people, they are objects of fascination, but also of mystery and superstition. Because their activity is restricted mostly to twilight and darkness and their days are spent in such places as old buildings, caves, and trees, they have been subject to misunderstanding and prejudice. They usually are seen, if at all, as darting, fluttering shadows in the dim light of evening or detected by the sound of moving wings and high-pitched chirps as they forage for insects in the night air. Researchers equipped with mist nets, head lights, electronic listening devices, and a willingness to explore damp or musty caves have begun to dispel many of our misconceptions and to reveal bats to be remarkable, uniquely adapted mammals.

The Chiroptera is a highly successful and diverse group, second only to rodents in number of species. Bats occur throughout the world, except in nonforested polar regions, but are most abundant in the tropics. Their diversity is reduced in temperate regions; nevertheless, 17 species (in 2 families and 8 genera) occur in the Carolinas, Virginia, and Maryland.

The most apparent structural adaptation of bats is modification of the forelimb to form a wing. Fingers 2 through 5 are extremely elongate and delicate; a thin membrane of skin extends between these fingers in a manner similar to fabric extending between the ribs of an umbrella. The membrane stretches from shoulder to finger tips and along the side of the body and the length of the hind leg at least to the ankle. An additional membrane, the interfemoral membrane, lies between the hind legs, usually enclosing the tail; the length of the tail and the extent of this membrane varies widely in different species. A cartilaginous projection called the calcar extends from the inside of each hind foot and provides additional support for the interfemoral membrane. Claws, which are present on the reduced first digit of the hand and on all the digits of the foot, enable bats to cling to their roosts during the day or when hibernating. Locomotion by means other than flight is accomplished only with some difficulty.

Vision is poorly developed in most bats, and they navigate on their nightly forays primarily by use of echolocation. Ultrasonic sounds are emitted through the mouth or nose, and returning echoes, received by the ears, allow the animals to determine such information as location, size, distance, and speed of nearby objects. Insect-eating bats locate food on the wing with this sophisticated mechanism and capture the insects by mouth, snare them in some part of the membranes, or pounce on them on the ground. Many bats have elaborate fleshy flaps, lobes, or projections of various and distinctive forms associated with the nose and/or ears, which apparently aid in the emission and reception of sound. One of the lobes associated with the ear, the tragus, refines the returning sound waves by intercepting such static as the impulses of other bats.

The bats of the Carolinas, Virginia, and Maryland are insectivorous, and their food supply tends to be unavailable during the winter months. Lowered environmental temperatures in the winter also place a great strain on active mammals, for they must increase their metabolism to maintain body heat. Therefore, the bats occurring in this region either hibernate, roosting alone or in colonies, or migrate to warmer latitudes.

Many species that live in temperate areas and hibernate in winter breed in the fall. Sperm typically are stored within the female reproductive tract until the following spring. Eggs then are fertilized by the sperm retained over winter. Such delayed fertilization many months after breeding varies significantly from the reproductive procedure common in most mammal groups. Infant bats tend to be large relative to the size of adults, perhaps weighing up to half as much as the mother. In some species the mother carries the young for several nights after birth on feeding flights rather than leave them behind in nursery areas of the colony; these females generally have large wings and strong flight capabilities.

Bats affect humans positively by their destruction of vast numbers of insects, seed dispersal by tropical fruit-eating species, and the pollination of important crops by nectar- and pollen-eating species. Bat guano has been a valued fertilizer in many parts of the world. On the other hand, several diseases are known to be carried by bats. Rabies, a viral infection of the central nervous system which is nearly always fatal, has received the most attention, but very few human cases in the United States have been bat transmitted. Another potentially harmful disease to humans is histoplasmosis, a fungus common in the southern United States which favors soil enriched by bird or bat droppings. Other diseases generally are restricted to the tropics and are of little importance to public health in our region. Any bat which exhibits unusual behavior should be reported to local health authorities, and handled with extreme caution.

Little Brown Myotis
Myotis lucifugus

Description. The long, glossy, dark brown fur and long hairs on the hind feet, which extend noticeably beyond the tips of the claws, best identify this relatively small bat. No single character, however, immediately distinguishes it from other species of *Myotis* in the region. The little brown myotis measures 3⅛ to 3⅞ inches (80 to 98 mm) in total length, and weighs ⅛ to ⅜ ounce (4 to 9 gm); the forearm is 1⅜ to 1⅝ inches (35 to 42 mm) long. The calcar lacks a keel, and the wing membrane attaches to the base of the toe. The face, ears, and wing membranes are dark brown.

The little brown myotis is confused most frequently with other species of

Little brown myotis (Myotis lucifugus).

Myotis, the big brown bat, and the evening bat. Means by which they can be separated are discussed in those accounts.

Distribution and Abundance. This bat is locally abundant in Maryland and Virginia, where several hundred individuals may occupy a single roost. It is uncommon and restricted to widely scattered roost sites in the piedmont and mountains of the Carolinas and apparently does not occur in the sandhills and coastal plain of these states except as an occasional vagrant.

Habitat. Attics of houses and buildings serve as maternity roosts during the summer, whereas single males have been found behind shutters and siding or under shingles and the loose bark of trees. Winter roosts invariably are caves and mines. Roosts are usually in close proximity to permanent bodies of water; this bat forages over ponds, rivers, and streams, or among the trees in broken forests.

Natural History. Maternity colonies form in May and consist primarily of females; at this time males roost alone or in small numbers away from the females. Individuals begin to migrate in midsummer, often flying several hundred miles to winter roosts. Males and females swarm around the entrances to these roosts in late summer and early fall, and by early winter most are hibernating.

The little brown myotis breeds in the fall and produces a single litter each year in late May or early June; a litter typically consists of a single offspring, but twins have been reported. The young are weaned and begin to fly at 3 weeks of age, and they reach sexual maturity in about 8 months. This bat may live as long as 24 years in the wild; males apparently live longer than females.

Moths, beetles, and other flying insects are captured in the flight membranes and then eaten. In autumn the little brown myotis acquires a thick layer of fat in preparation for the period of hibernation. Bats in this region consume a tremendous amount of insects; for example, a popular magazine recently reported that bats consume an average of 13 tons of insects each year in a city the size of Boston, Massachusetts. The little brown myotis is known to carry rabies, but the incidence of this disease is less than 1 percent.

Southeastern Myotis
Myotis austroriparius

Description. The southeastern myotis is most similar to the little brown myotis in appearance, but the dull, woolly fur

Southeastern myotis (Myotis austroriparius).

and pink face distinguish it from that species. The southeastern myotis is relatively large, measuring 3¼ to 3¾ inches (84 to 96 mm) in total length, and weighing about ⅛ to ¼ ounce (5 to 7 gm); the forearm is 1⅜ to 1⅝ inches (36 to 42 mm) long. Females are slightly larger than males. The fur is not distinctly banded, being grayish brown above and white or tan below; occasional individuals are yellowish brown. The calcar is not keeled, the wing membrane attaches to the base of the toe, and the long hair of the feet extends well beyond the tips of the toes.

Distribution and Abundance. There are only 3 records of this relatively rare bat from this region. They are from the piedmont (Wake County) and coastal plain (Pender County) of North Carolina and the coastal plain (Kershaw County) of South Carolina.

The Wake County bats inhabited the understory of an abandoned mill along the Neuse River, which has since been converted into condominiums, and the Pender County record was of a single specimen taken from a dock piling on the Cape Fear River. The South Carolina specimen

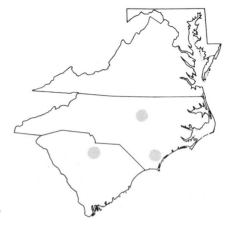

came from a factory in Camden near the Catawba River. We frequently see them darting over the Northeast Cape Fear River near Wilmington, North Carolina.

Habitat. In other parts of its range, caves usually serve as daytime roosts, especially during warmer months; buildings, hollow trees, and sewers are favored winter roosts. Roost sites are always near rivers or other permanent bodies of water.

Natural History. Individuals congregate by the hundreds or thousands in late spring in maternity roosts, either caves or the cellars of buildings that usually have standing water on the floor of the site. Pregnant females isolate themselves from other bats and form a nursery colony in a well-protected part of the cave or perhaps under the siding of a building. Each female gives birth to 2 (occasionally 1) offspring in May. The young begin to fly in 4 to 5 weeks, and shortly thereafter they join the summer colony. Mating peaks in fall, but continues until spring.

Individuals begin to leave the summer colony as cold weather approaches, and by December most bats have departed; however, some animals, usually young of the year, remain throughout the winter. Winter roosts are typically boathouses, crevices in docks and bridges, or hollow trees and are usually located over water. Here the animals cluster in small numbers and hibernate. Hibernation is broken by periodic bouts of activity in which each bat flies about to drink, defecate, and possibly mate.

Other aspects of the natural history of the southeastern myotis are probably similar to those of the little brown myotis. Insects are taken over rivers, lakes, or other permanent bodies of water.

Gray Myotis
Myotis grisescens
(Endangered)

Description. The peculiar attachment of the wing membrane to the ankle rather than to the base of the toe distinguishes the gray myotis from other species of bats in the region. The unbanded fur of this bat is gray or dusky above, but frosted with white on the belly. The calcar is not keeled. The gray myotis is relatively large, measuring 3⅛ to 3¾ inches (80 to 96 mm) in total length, with a forearm length of 1⅝ to 1¾ inches (41 to 46 mm), and weighing ¼ to ⅜ ounce (6 to 9 gm) in specimens from the central Mississippi River Valley.

Distribution and Abundance. The stronghold of this Endangered bat is Missouri, Kentucky, Tennessee, Alabama, and adjacent states where it is extremely habitat specific and declining in numbers. Records from this four-state region are from the mountains of western North Carolina (Buncombe County) and Virginia (Lee and Scott counties). The North Carolina record, a single animal which had been tagged in Tennessee, probably was a vagrant. The Virginia records are of 3 summer bachelor colonies, each containing several hundred individuals.

Gray myotis (Myotis grisescens). *Photograph by John R. MacGregor.*

Habitat. The gray bat inhabits relatively large limestone caves, on the floor of which run small rivers and creeks. Only a half-dozen caves are known to serve as winter hibernacula, and none are protected from human disturbance.

Natural History. Little is known about the gray bat in this region. In the central Mississippi Valley, it roosts the entire year in caves, migrating up to 300 miles (484 km) between winter hibernacula and summer maternity or bachelor quarters. Thousands of females form maternity colonies in the early spring, while males congregate in smaller groups in other caves. Each female gives birth to 1 offspring in late May or early June. The young begin to fly in July, but do not become sexually active until 2 years of age.

Individuals depart for winter roosts in late summer and early fall and mate there during October and November. They hibernate by the thousands when cave temperatures reach 45° to 50°F (7° to 10°C), frequently forming clusters several tiers deep.

Roosts are always near large bodies of permanent water such as rivers and reservoirs, over which this bat forages at night for mayflies and other flying insects.

Keen's Myotis
Myotis keenii

Description. The large ear, which when laid forward extends well beyond the muzzle, and the long pointed tragus separate Keen's myotis from other species of *Myotis* that occur in the four-state region. Keen's myotis is 3 to 3⅞ inches (75 to 100 mm) in total length, its ear measures ⅝ to ¾ inch (17 to 19 mm), and its forearm is 1⅜ to 1½ inches (35 to 39 mm) long; it weighs ⅛ to ¼ ounce (5 to 7 gm). The calcar is not keeled, and the dull brown fur is slightly paler below than above.

Distribution and Abundance. This bat is uncommon throughout Maryland, Virginia, and the mountains of the Carolinas. There is a single record from the piedmont (Wake County) of North Carolina.

Habitat. Caves and mines serve as night roosts, but during the day Keen's myotis is found most often in buildings and hollow trees or under shutters, shingles, and the loose bark of trees. Roosts are usually found in heavily forested areas.

Natural History. Animals typically roost singly or in small maternity groups but never in the large colonies typical of many other species of bats in the four-state region. Sometimes Keen's myotis roosts with the little brown myotis, but, in general, the former prefers to roost separately in colder enclaves of the roost.

Little is known about the reproductive biology of Keen's myotis. It swarms around entrances to caves and

Keen's myotis (Myotis keenii). *Photograph by John R. MacGregor.*

as it is nowhere abundant and seldom inhabits buildings in large numbers.

Indiana or Social Myotis
Myotis sodalis
(Endangered)

Description. This relatively rare bat can easily be mistaken for other species of *Myotis*, and a careful consideration of several characters is necessary to identify it positively. The hair on the feet is short and does not extend beyond the tips of the claws, the tail membrane attaches to the base of the toe, and the calcar is keeled. The dull, wooly fur is distinctly banded, the basal portions of each hair being darker than either the grayish middle band or the cinnamon-brown tip, resulting in an overall color of brownish

mines in August, a ritual which apparently serves for the recognition of potential mates. Females give birth to a single offspring sometime in July.

This bat forages late in the evening for small soft-bodied insects. It is of little economic importance to humans

Indiana or social myotis (Myotis sodalis).

to grayish black on the back and pinkish to cinnamon on the belly. The nose is pink to pale brown, and the wing membranes are blackish brown. The Indiana myotis measures 3 to 3⅝ inches (77 to 91 mm) in total length, and its forearm is 1⅜ to 1⅝ inches (36 to 41 mm) long; it weighs ⅛ to ¼ ounce (5 to 8 gm).

Distribution and Abundance. Records of occurrence are limited to western North Carolina, Virginia, and Maryland. The Indiana myotis is extremely localized throughout its geographic range and has undergone a drastic reduction in numbers in recent years. Its roost requirements are extremely specific, and only a few caves and mines are known to serve as winter hibernacula. Disturbances greatly affect the total numbers in a roost; for example, a flood in March 1964 killed approximately 85 percent of a cave population in Kentucky. The Indiana myotis is included on the federal list of Endangered Species.

Habitat. Hibernacula are usually limestone caves with standing water on the floor of the roost. During summer males apparently roost in small numbers in caves, and females form maternity colonies in hollow trees and underneath loose bark. Most summer roosts are near streams or small rivers.

Natural History. The Indiana myotis forages at night for moths, mayflies, and other flying insects near the tops of trees and over streams. They hibernate from early October until late April in compact clusters of 500 to 5,000 bats. Individuals awaken periodically (8 to 10 days) and move to warmer sections of the cave where other alert bats have congregated; they then return to a hibernating cluster and again become torpid.

Migration occurs in early September as bats fly as much as 300 miles (500 km) to winter roosts. Mating occurs at night in early October, and a single young is born in late June or early July; it begins to fly at about 1 month of age. The Indiana myotis is long lived, some specimens reaching at least 20 years of age.

Small-footed Myotis
Myotis leibii

Description. This is the smallest member of the genus occurring in the region, measuring 2⅞ to 3¼ inches (73 to 82 mm) in total length; the forearm is 1¼ to 1⅜ inches (31 to 34 mm) in length, and the hind foot is ¼ to ⅜ inch (7 to 8 mm) long. Its fur is chest-

Small-footed myotis (Myotis leibii).

nut brown above and grayish brown below. The small ears and face are black, giving the animal a masked appearance, the flight membranes are blackish brown, and the calcar is keeled.

The small-footed myotis is confused most frequently with the little brown myotis; however, the small-footed myotis is smaller in size, has a black face, a keeled calcar, and the fur is not as dark.

Distribution and Abundance. This bat occurs in the Appalachian Mountains of western Maryland, Virginia, and North Carolina, where it is uncommon; there also is a record of its presence in the piedmont (Montgomery County) in Maryland. Within this part of its range, the small-footed myotis is found most often near caves, but is rarely encountered elsewhere.

Habitat. Little is known about the habitat requirements of this species in the southern Appalachians, although roosts usually are found in hemlock forests. It roosts during the fall and

winter in inconspicuous places such as under boulders, in crevices in rock falls and quarries, or around the entrances to caves and mines. It may frequent buildings in the summer.

Natural History. The small-footed myotis hibernates only during the coldest periods in winter and early spring, being the last cave-dwelling species of *Myotis* to enter torpor and the first to become active in the spring. While hibernating, the small-footed myotis hangs with its forearms outstretched from the body at a 30 degree angle rather than parallel to the body as do most other species of bats.

Virtually nothing else is known about the biology of the small-footed myotis in this region. Based on the patterns of reproduction in other species of *Myotis*, females probably give birth to a single young in May or June following the fall and winter breeding period.

Silver-haired Bat
Lasionycteris noctivagans

Description. The blackish brown hair of this handsome bat is tipped with silver, especially on the shoulders and back. The interfemoral membrane is moderately furred above, especially near the body, and the ears and tragi are short, blunt, and naked. The face, ears, feet, and flight membranes are blackish brown. This is a medium-sized bat, measuring 3¾ to 4⅝ inches (96 to 116 mm) in total length, with a forearm length of 1⅝ to 1¾ inches

Silver-haired bat (Lasionycteris noctivagans).

(40 to 44 mm); it weighs ¼ to ⅜ ounce (8 to 11 gm). Juveniles have more silver-tipped hairs than do adults, but males and females of the same age appear identical in color and size.

The silver-haired bat and the hoary bat might be confused because both have dark fur with silver-tipped hairs. However, the hoary bat is much larger in size, it has a densely furred interfemoral membrane and distinct patches of fur on the wrist and elbow and inside the ear, and it is more brownish in color.

Distribution and Abundance. There are records of this relatively uncommon bat from Maryland, Virginia, and the Carolinas, but it occurs sporadically both in distribution and abundance. This species is migratory, with most individuals moving northward in the spring and returning in the fall. Males, however, do not migrate as far north as females; for instance, all summer records from South Carolina are of males. There are no summer records from Virginia.

Habitat. These bats prefer to roost near permanent water in clumps of leaves, abandoned woodpecker holes, and protected crevices under loose bark in trees. Rock crevices and relatively open buildings sometimes serve as daytime roosts. The most secure of these sites are used as hibernacula during the coldest periods of winter.

Natural History. Many aspects of the biology of the silver-haired bat are poorly known inasmuch as it does not roost in large congregations or in attics of houses as do many other species of bats. Most information has come from specimens taken during spring and autumnal migrations or shot while flying over permanent bodies of water.

Males and females copulate in the fall, but fertilization does not occur until spring. Females form maternity colonies in late spring and typically give birth to 2 offspring in June. Most of the young are weaned and on the wing by late July and are apparently capable of reproducing by their first autumn. Males remain solitary throughout much of the year.

In this region, the silver-haired bat emerges an hour or so after sunset to feed from ground to treetop level over ponds and streams in forested areas. It flies alone or in small groups in a relatively straight and lethargic manner, usually making several passes over the same area during the night.

Some bats of this species have been found to be rabid. Most of these are the young of the year, usually taken in August.

Eastern Pipistrelle
Pipistrellus subflavus

Description. This is the smallest of all bats occurring in the four-state region, weighing only about a fifth of an ounce (6 gm) and measuring 2¾ to 3¾ inches (71 to 95 mm) in total length; the forearm is 1¼ to 1⅜ inches (33 to 36 mm) in length. Females are slightly larger than males; both sexes are heavier in the fall from fat accumulation. The fur of this bat is tricolored; each hair has a dark brown base, a pale middle band, and a dark tip, generally resulting in a overall yellowish brown hue. The wing membranes are blackish brown, and the face and ears are pale brown. The tragus is blunt. That part of the interfemoral membrane nearest the body is furred. The small size, blunt tragus, and tricolored fur separate the eastern pipistrelle from other species of bats that occur in the region.

Distribution and Abundance. The eastern pipistrelle occurs throughout the Carolinas, Virginia, and Maryland. It

is one of the most common bats in the piedmont and mountains and is locally abundant in the coastal plain.

Habitat. Spanish moss and clumps of leaves are used as daytime roost sites during much of the year; caves, rock crevices, and mines serve as hibernacula and summer night roosts. Small maternity colonies sometimes are in relatively open houses and buildings and are often exposed to more daylight than most other species of bats would tolerate.

Natural History. As dusk approaches, the eastern pipistrelle begins its erratic flight, which has been likened to that of a fluttering moth. It flies at treetop level over permanent bodies of water and among scattered clumps of trees, foraging for moths and other flying insects. It forages alone during much of the year, but in the late summer congregates in greater numbers around the mouths of caves and favored feeding places.

The gestation period is not known, but this species breeds in November, and the sperm remain viable until April when fertilization probably occurs. Usually 2 young are born in June, their combined weight being about one-third that of their mother. They grow quickly and can fly in about 1 month. Some males live 15 years, but females seldom live to reach 10 years of age.

The eastern pipistrelle hibernates in small caves or small protected arms of larger caves where winter temperatures remain relatively constant. They typically hang alone but, if space permits, several hundred may hibernate in the same cave. Each individual

Eastern pipistrelle (Pipistrellus subflavus).

awakens periodically to fly and void waste material; afterwards it returns to one of several favorite locations in the hibernaculum and again becomes torpid. Droplets of water condense on the fur if the bat remains in one position for an extended length of time. Females emerge from hibernation before males.

Big Brown Bat
Eptesicus fuscus

Description. The big brown bat has long, glossy brown fur and is relatively large in size, measuring 4 to 5⅛ inches (103 to 130 mm) in total length and weighing ½ to ⅝ ounce (13 to 18 gm); the forearm is 1¾ to 1⅞ inches (45 to 48 mm) long. Females average larger than males. The flight membranes are slightly darker in color than the fur, the calcar is distinctly keeled, and the tragus is broad and rounded.

Several species of *Myotis* and the evening bat superficially resemble the big brown bat, but they are distinctly smaller in size and either lack a keeled calcar and/or possess a pointed tragus.

Distribution and Abundance. While there are records of the big brown bat from every state in the region, it is uncommon throughout most of the area and rare in the coastal plain of eastern North Carolina. It is familiar to humans because it frequently finds its way into rooms from between walls, the attic, or alongside the chimney.

Habitat. This bat favors buildings, and roosts in relatively large colonies in attics or in smaller groups behind window shutters and under eaves. Hollow trees and crevices in rocks and under loose bark also are utilized to a lesser extent.

Natural History. During the coldest part of the year, usually from December until March, this permanent resident hibernates in well-protected roosts such as caves, mines, and buildings. Air temperatures determine the depth of torpor, however, and on warm winter afternoons individuals awaken to drink, defecate, and breed.

Mating occurs intermittently from November until March. The pregnant females then segregate themselves, forming a maternity colony. Usually 2 (range 1 to 4) young are born in late May or early June after a gestation period of about 2 months. The young are on the wing by late June, and many become reproductively active during their first year. This bat is long lived, and individuals are known to live at least 18 years.

Big brown bat (Eptesicus fuscus).

The big brown bat consumes beetles, wasps, and most other flying insects, some of them pests to humans; moths are seldom eaten. It forages nightly over streams and grasslands, in broken forests, and in cities. A thick layer of fat, acquired in the fall, provides a necessary source of energy to allow metabolic processes to continue during hibernation.

The big brown bat, and probably other species of bats as well, is extremely sensitive to DDT and related pesticides. Because these pesticides are fat soluble, individuals with a thick layer of fat in autumn are affected little; however, the lean bats of spring are poisoned when they consume insects with pesticides in their tissues. The big brown bat also is known to carry rabies.

Red Bat

Lasiurus borealis

Description. The red bat is easily distinguished from other bats in the region by its unique color and completely furred tail membrane. The silky dorsal fur is brick to rusty red, whereas the belly is slightly paler; males are more brightly colored than females. The ears are short and rounded, but the wings and tail are long and pointed. This moderately sized species is indistinguishable from the Seminole bat in size; selected external measurements of both are included in the account of the latter. Female red bats average slightly larger than males in body size.

Distribution and Abundance. This is the most common and widespread bat in the region; it is known from both barrier islands and mountaintops. It is migratory, but local patterns of migration are not known.

Habitat. Favored roost sites are trees and shrubs, often near permanent water or open fields. Individuals usually hang from south-facing twigs and stems in such a manner that they are protected from the sun, elements, and predators. Most roosts are 4 to 10 feet (1.2 to 3 m) above the ground, but some are near or on the ground; families and young individuals tend to roost from 10 to 20 feet (3 to 6 m) high. Red bats also have been captured near the mouths of caves in July, August, and September.

Natural History. Because the red bat is so abundant in the region, much is known about its biology. Except for a mother and her young, individuals remain solitary most of the year. Copulation occurs in August and September, often beginning in flight. One to

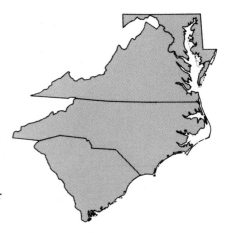

Red bat (Lasiurus borealis), *female.*

Red bat (Lasiurus borealis), *male.*

5 young are born in late May or early June after a gestation period of 80 to 90 days. The young are born hairless and with their eyes closed, but grow rapidly and are capable of flight in 3 to 6 weeks. Infant mortality is high until early August, when the young are weaned.

Red bats begin to forage shortly after sunset over streams, fields, pastures, or around street lights in suburban areas. Routinely, the red bat flies in graceful arcs with the tail extended straight behind the body; however, flight becomes erratic whenever it chases a potential meal of moths, flies, beetles, crickets, bugs, and other insects. Predators of red bats include opossums, cats, and various species of hawks and owls; also, blue jays are known to consume this bat, especially young individuals.

As winter temperatures begin to drop, the red bat conserves energy by wrapping its heavily furred tail over its belly, chest, and wings, and hibernates. It may awaken on warm winter days to forage, usually before dusk, if prey is available.

The red bat is encountered most often by humans under trees where it sometimes falls from the combined weight of the nursing young. It is known to carry rabies, so individuals should be handled with caution.

Seminole Bat
Lasiurus seminolus

Description. The Seminole bat is similar in appearance to the red bat; they are indistinguishable in body size and both possess a densely furred tail membrane and short, rounded ears. The fur of the Seminole bat, however, is deep mahogany in color, with a slight frosting of white, whereas the fur of the red bat is reddish. The wing membranes of the Seminole bat are dark and there is a prominent patch of white fur on each shoulder and wrist. Seminole and red bats measure 3½ to 4¾ inches (88 to 121 mm) in total length, and weigh ¼ to ½ ounce (7 to 14 gm); their forearms are 1¼ to 1¾ inches (32 to 44 mm) long. Female Seminole bats average slightly larger than males.

Distribution and Abundance. This relatively common bat is a denizen of the lower piedmont and coastal plain of the Carolinas, but the northern limit of its range in the region is poorly documented. Evidently it is most common during the summer months, but we have collected specimens in southeastern North Carolina during the winter. Females appear to be more

Seminole bat (Lasiurus seminolus). *Photograph by John L. Tveten.*

common during the summer than males; males are more numerous during other seasons of the year.

Habitat. The Seminole bat prefers to roost in clumps of Spanish moss or other dense clumps of foliage, preferably 3¼ to 15 feet (1 to 4.5 m) above the ground on the southwest side of trees. Roosts are open from below, and the ground beneath the roost is covered with leaves, which minimizes reflected sunlight. Roosts are frequently situated near open areas over which feeding occurs. As with other species of bats that seek shelter in trees, they tend to be solitary.

Natural History. Shortly after dusk the Seminole bat begins to forage near open water and over clearings in pine barrens and hummocks for flies, beetles, and other flying insects. Even ground-dwelling crickets have been consumed. Seminole bats also frequent street lights to feed on insects attracted to the light. This bat becomes torpid during the colder months, but on warm evenings during the winter it may become aroused long enough to feed if prey are available.

Little is known about the breeding biology of the Seminole bat. Birth probably occurs in late May and June; 1 to 4 embryos have been recorded. Evidently females find alternative roosts in which to give birth and raise their young because few families have been found in clumps of Spanish moss. The young are probably on the wing in 3 to 4 weeks.

The economic importance of this species is negligible in the region

owing to its restricted and seasonal distribution. Much remains to be learned of its biology, including the northern limit of its winter range, seasonal patterns of movement, and breeding biology.

Hoary Bat
Lasiurus cinereus

Description. The hoary bat is quite distinctive, being the largest bat in the region and having a heavily furred tail membrane. The fur is distinctly banded; the hairs are black basally, then tan, dark brown, and frosted white at the tips, giving the animal a hoary appearance. The fur on the throat and under the forearm is yellowish, and patches of white fur are present on the shoulders and wrists. The short, rounded ears are edged in black, and the keeled calcar has distinct lobes on the tip. This bat can be identified on the wing by its swift flight and large size. It is 4⅞ to 5⅝ inches (123 to 142 mm) in total length, and weighs ⅝ to 1 ounce (19 to 28 gm); the forearm measures 2 to 2⅛ inches (52 to 54 mm) in length. Females are slightly larger than males.

Distribution and Abundance. This bat is uncommon throughout Maryland, Virginia, and the Carolinas. It migrates northward during the warmer months, and no specimens have been taken in summer in Virginia or the Carolinas.

Habitat. Individuals apparently roost separately during the day in forested

Hoary bat (Lasiurus cinereus).

habitats, particularly in coniferous forests that border cleared areas and permanent water over which they feed. Roost sites are usually 10 to 15 feet (3 to 4.5 m) above the ground in clumps of dense foliage. Roosts open at the bottom for easy movement in and out.

Natural History. The sexes appear to be segregated during most of the year. Breeding takes place during autumn and the young are born from mid-May until early July; litter size varies from 1 to 4, but 2 is typical. Newborn bats are covered with short hair on the back, but the belly is naked; the ears and eyes are closed. The offspring grow rapidly and are capable of flight in 4 to 5 weeks.

Groups of hoary bats have been observed to migrate in spring and autumn. In northern Florida, departure to the northerly breeding range occurs from February to May; autumnal movements to the south are in October and November. Unfortunately, there is little information concerning migratory patterns in this region.

Northern populations may hibernate during colder months rather than migrate, probably using the tail membrane for a blanket as does the red bat.

Moths are preferred as food but beetles, flies, dragonflies, grasshoppers, termites, wasps, and even smaller bats are also consumed. The stomach of one hoary bat even contained grass and bits of snake skin. Several individuals may forage in a group, generally emerging well after sunset; however, they may also forage in the late afternoon on warm winter days.

The hoary bat is uncommon and of little economic importance in the region. It is rarely encountered by humans because it seldom roosts in houses and other buildings. It is known to carry rabies, however, and should be handled with extreme care.

Northern Yellow Bat
Lasiurus intermedius

Description. This is the only large bat in the region in which the tail membrane is furred only on the anterior half of the dorsal surface. The long, silky pelage is yellowish orange to yellowish brown and sometimes faintly washed with black. The wing membranes are brownish and the face and ears are pinkish brown. The calcar is slightly keeled. This bat measures 4⅝ to 5⅛ inches (118 to 129 mm) in total length; it weighs ½ to ⅝ ounce (15 to 19 gm), and the forearm is 1⅞ to 2 inches (48 to 51 mm) long. Females

Northern yellow bat (Lasiurus intermedius). *Photograph by John L. Tveten.*

average larger than males in most body measurements.

Distribution and Abundance. The northern yellow bat appears to be restricted in this region to the southern Atlantic seaboard; all records are from coastal South Carolina except for a single specimen from Willoughby Beach, Virginia. In that Spanish moss, its favorite roost, is continuously distributed at least as far north as the tidewater areas of eastern Virginia, it is possible that the northern yellow bat occurs throughout coastal North Carolina. The specimen from Virginia carried 3 embryos, indicating that a permanent breeding population may exist there. This bat is rare in this part of its range.

Habitat. Clumps of Spanish moss frequently are used as daytime roosts, particularly in stands of long-leaf pine and turkey oak. Wooded areas in the vicinity of permanent water and large open areas, over which this bat feeds, are preferred. The northern yellow bat is seldom associated with buildings or caves.

Natural History. The northern yellow bat forages at night at heights of 15 to 20 feet (4.5 to 6 m) over grassy areas or along forest edges for flies, mosquitos, and other flying insects. Large numbers of females and young form feeding aggregations in the summer. The sexes are segregated during the summer; the males are solitary, whereas several females may form a loosely knit colony in a single moss-covered tree. Males apparently congregate in the winter, but little is known of female roosting habits during this time of the year. The northern yellow bat becomes torpid when exposed to cool temperatures in the northern part of its range.

The reproductive pattern is not well known for this species. Mating apparently occurs in autumn and winter, 2 to 4 young are born in late May or June, and the young are on the wing by late June or early July. Females do not carry their young on nocturnal feeding flights, but may transport their offspring from daytime roosts when disturbed.

Because the northern yellow bat is rare in the region, it is encountered infrequently. It also is known to carry rabies.

Evening Bat
Nycticeius humeralis

Description. This relatively small brown bat is nondescript and often confused with the big brown bat and many species of *Myotis*. However, the evening bat is distinctly smaller than the big brown bat, and the shape of the tragus (blunt and curved forward) separates the evening bat from all species of *Myotis* (long and pointed). The tail of the evening bat is not furred, and the short ears and wing membranes are blackish brown and leathery. The calcar is not keeled. The fur is short and dull, dark brown above and paler below. Juveniles are noticeably darker than adults in pelage coloration until about 6 weeks of age when the first molt occurs. This bat measures 3 to 3⅞ inches (75 to 99 mm) in total length, and weighs ¼ to ½ ounce (6 to 13 gm); the forearm is 1¼ to 1⅝ inches (31 to 40 mm) long. Females average larger than males in most dimensions.

Distribution and Abundance. The evening bat is distributed at lower elevations throughout the region, and is not known from the mountains of Virginia or Maryland. It is relatively abundant during the spring and summer, but migrates southward for the fall and winter.

Habitat. This is a woodland species that usually roosts in hollow trees or in crevices under bark; it also has been collected in buildings and under bridges, the buildings frequently being shared with Brazilian free-tailed bats. It may swarm around the mouths of caves and mines in the autumn, but it does not inhabit them.

Natural History. As its name implies, the evening bat is nocturnal, emerging well after dusk to search for food. It forages in a slow and steady flight early in the evening. Stomach contents of only a few individuals have been examined, so little is known about its feeding habits. Evidently, moths, beetles, flies, and flying ants are consumed.

The sexes are separated during the spring when pregnant females form roosting clusters consisting of a few to several hundred individuals, usually in the attics of buildings. From 1 to 4 embryos have been found, but usually 2 young are born in late May or June. Little is known of the male reproductive cycle or when copulation occurs, but both sexes have been taken together in August and breeding may occur then. Young are naked at birth but grow rapidly; they begin to fly in 3 weeks and are weaned between 6 and 9 weeks of age.

The evening bat accumulates large reservoirs of fat in the autumn, and hibernates during the colder months in the Deep South. Considering its relative abundance, a scant amount of information is known about this species.

Evening bat (Nycticeius humeralis).

Townsend's Big-eared Bat

Plecotus townsendii
(Endangered)

Description. Both Townsend's and Rafinesque's big-eared bats are easily distinguished from other species of bats by their extremely large ears, which, when laid back, are approximately half the length of the body. The ears are thin, furred only along the edges, and joined across the forehead at their bases; they often are coiled against the body during rest or torpor. These bats are sometimes called lump-nosed bats due to a conspicuous fleshy growth on either side of the muzzle between the nostril and eye.

The fur of Townsend's big-eared bat is long and soft, and there is little contrast between the bases and tips of the hairs. The result is an overall brownish color that is slightly darker above than below. The hair on the feet does not extend beyond the tips of the toes. Animals from West Virginia have a total length of 3¾ to 4⅜ inches (96 to 112 mm); the forearm measures 1⅝ to 1⅞ inches (42 to 47 mm), and the weight is about ⅜ ounce (9 to 12 gm). Females are larger than males in some dimensions, most notably increased body weight during the fall and winter.

Distribution and Abundance. Although common in western North America, it is rare in this region, being represented by an isolated race that occurs in the mountains of western Virginia and North Carolina. This race is included on the federal list of Endangered Species because numbers have decreased in recent years as a result of disturbances by spelunkers, vandalism, and loss of suitable habitat.

Habitat. Limestone caves in mountains over 1,500 feet (460 m) in elevation are utilized as daytime roosts. Individuals congregate in small groups along the ceilings and walls, usually near the mouth of the cave. Townsend's big-eared bat apparently uses the same maternity roosts year after year and hibernates (rather than migrates) during the winter. Hibernation, however, is broken by periodic bouts of activity, when individuals move within the hibernaculum or to nearby caves.

Natural History. As dusk approaches, Townsend's big-eared bat makes several passes to the opening of the roost as if checking the amount of available light, only to remain inside the roost and wait for darkness to intensify. These bats exit the roost well after sunset to forage for moths, beetles, flies, wasps, and winged ants.

Mating occurs in fall and winter,

Townsend's big-eared bat (Plecotus townsendii).

and a single young is born in June. Yearling females are reproductively active, but males do not enter the breeding population until their second autumn. Females form maternity colonies during the spring and summer; males are usually solitary at this time. In winter, however, individuals roost alone or form small clusters comprised of both sexes.

Rafinesque's Big-eared Bat
Plecotus rafinesquii

Description. Rafinesque's big-eared bat closely resembles Townsend's big-eared bat, but is distinguished from the latter by its long bicolored fur. Individual hairs are blackish at the base, with those on the back having yellowish brown tips and those on the underside having whitish tips; the animals therefore appear grayish brown above and silvery below. Long hairs on the feet extend well beyond the tips of the claws; the wing and tail membranes are thin and naked. The total length ranges from 3¾ to 4⅛ inches (95 to 105 mm), with a forearm of 1½ to 1¾ inches (38 to 44 mm); this animal weighs ¼ to ⅜ ounce (6.0 to 9.5 gm). Females are heavier than males, but the sexes are similar in other body dimensions.

Distribution and Abundance. Rafinesque's big-eared bat is uncommon to rare in the region, being restricted to the mountains of western North Carolina and the sandhills and coastal plains of the Carolinas as far north as the Dismal Swamp in southeastern Virginia. Specimens from the mountains tend to be paler in coloration than those from the coast.

Habitat. This species prefers to roost in dilapidated houses and buildings near permanent water; however, specimens have been found in hollow trees, behind loose bark, or at the entrances to caves and mines. Most roosting sites are partially lighted and relatively dry.

Natural History. Males and females inhabit the same roost during late fall and early winter when breeding occurs. During spring, however, females form maternity colonies consisting of several to approximately 100 bats, whereas males are solitary or congregate in small numbers away from females. A single naked young, born in late May or early June, is capable of flight at about 3 weeks of age and is similar in size to adults by 4 weeks of age. Individuals may live at least 10 years in the wild.

Rafinesque's big-eared bat is a per-

Rafinesque's big-eared bat (Plecotus rafinesquii).

manent resident in the region and hibernates during the winter and early
spring; it also may become torpid during summer cold spells. In torpor or
hibernation, as well as at rest during
the day, the ears are folded back next
to the body and under the wings. At
the least disturbance this bat becomes
alert, the ears become erect, and the
animal moves its head from side to
side; it quickly takes flight and is extremely maneuverable.

This bat is particularly susceptible
to human interference and does not
do well in captivity. Little is known of
its feeding habits, and many other aspects of its biology are poorly known
as well.

Brazilian Free-tailed Bat
Tadarida brasiliensis

Description. The Brazilian free-tailed
bat may be separated from other bats
in the region on the basis of a tail that
extends for about half its length beyond the naked interfemoral membrane, and short velvety fur. The
pelage of this bat is brownish gray in
color, but some animals have individual white hairs or patches of white
fur scattered randomly about. The
bases of the thick leathery ears are
separated narrowly at the middle of
the forehead, and the ears are directed more forward than those of
other species of bats in the region.
The wings are long and narrow, and

Brazilian free-tailed bat (Tadarida brasiliensis). *Photograph by Nicky L. Olsen.*

the hair on the feet extends well beyond the tips of the claws. The Brazilian free-tailed bat is 3½ to 4 inches (88 to 103 mm) in total length, and the forearm measures 1⅝ to 1⅞ inches (41 to 47 mm) in length; this bat weighs ⅜ to ½ ounce (9 to 13 gm). Males are slightly larger than females.

Distribution and Abundance. This bat is distributed throughout southern North America, and reaches the northern limit of its range along the Atlantic coast in widely scattered localities in the sandhills and coastal plain of the Carolinas. Its abundance in this area has not been well documented, but it is extremely gregarious, so large local populations should not be unexpected. Individuals in this region do not migrate.

Habitat. Buildings, schools, and other man-made structures are used as roost sites by Brazilian free-tailed bats. Some of these roosts are utilized throughout the year, whereas others are vacated during the winter by some or all of the summer residents.

Natural History. The sexes roost together during the fall, winter, and spring; however, in the summer pregnant females form large maternity colonies, while males segregate in smaller groups in other sections of the roost or scatter singly among the females. Mating occurs in February and March, and a single offspring is born in June following a gestation period of about 2 months. Newborn young are blind and naked at birth, but they are capable of flight at about 5 weeks of age. Females do not form strong bonds with their offspring; for example, lactating females indiscriminately allow any newborn to suckle. Reproductive capacity is attained within the first year.

Individuals become increasingly active as dusk approaches, then exit the roost in rapid succession. They fly to a preferred feeding site, frequently a permanent body of water many miles away, and forage much higher in the night sky than other bats for moths and other winged insects. The majority of these bats return to the roost shortly before dawn and do not use night roosts. The flight of the Brazilian free-tailed bat is more rapid than that of other bats in the region, and animals sometimes reach several thousand feet in altitude.

This species does not hibernate, but becomes lethargic when air temperatures drop in winter. The Brazilian free-tailed bat is known to harbor and transmit rabies; other infectious diseases such as histoplasmosis thrive in the accumulated feces, or guano, on the floor of the roost.

Armadillos
Order Edentata

There are 3 families of edentates, containing tree sloths, anteaters, and armadillos. All 3 occur in South America, but only the armadillos have extended their range of distribution into the United States. The term "edentate" means without teeth and is misleading, because only anteaters have no teeth; however, tree sloths and armadillos lack incisors and canines and the remaining cheek teeth are all alike, simple (peglike), and without enamel.

The family of armadillos (Dasypodidae) contains 9 genera and 20 species. It ranges from Argentina and Chile through the southern United States, but the nine-banded armadillo is the only member of the family that occurs here.

Nine-banded Armadillo

Dasypus novemcinctus
(Introduced)

Description. This animal is unique among North American mammals in that most of the body is covered by a protective armor of fused bony plates overlain by leathery skin. Small scales are present on the legs, but only soft skin covers the underside of the animal. The pelage is reduced to sparse hair on the belly and between the plates. The small head tapers to a soft, piglike snout; ears are long and naked. The legs are short and strong with stout claws. The carapace is yellowish to mottled dark brown. Total length in animals from Texas averages about 30 inches (76 cm); weight in males ranges from 12 to 17 pounds (5.4 to 7.7 kg) and in females, from 8 to 13 pounds (3.6 to 6.0 kg).

Distribution and Abundance. First recorded in the United States in 1854 from south Texas, the nine-banded armadillo has steadily extended its range northward and eastward. Its presence in southeastern United States was accelerated by introductions into Florida and the escape or release of captive animals. Reports of this armadillo from the piedmont of the Carolinas and Washington, D.C., probably represent escaped individuals. A specimen from Allendale County, South Carolina, may constitute the northernmost limit of its expanding range along the Atlantic seaboard, but the status in South Carolina remains uncertain. As it is essentially a tropical mammal, the northward extension of its range seems to be limited primarily by the severity of winter climate.

Habitat. The nine-banded armadillo may be found in a variety of habitats, including wooded forests or bottomlands, shrubby areas, and relatively open fields. It is a burrowing animal and, therefore, prefers loose soil to predominately clay or rocky soils.

Nine-banded armadillo (Dasypus novemcinctus). *Photograph by William F. Sanderson.*

Natural History. This armadillo is active primarily at dusk and at night, especially in summer; in midwinter it usually emerges from its burrow in the early afternoon or the warmest part of the day. It forages for food under cover in daylight and moves into more open areas at night. Over 90 percent of its diet consists of insects and other invertebrates; the remainder is made up of amphibians, reptiles, berries, fruit, and roots. Its weak jaws and peglike teeth dictate that food be relatively soft. It utilizes a keen sense of smell to locate food, which it digs from the soil with the powerful action of the forefeet and claws.

Burrows are dug under stumps, logs, bushes, and brush piles, in stream banks, on hillsides, and other places with adequate protective cover. Most burrows lead to an enlarged chamber, typically located about 1½ feet (0.5 m) below ground surface, containing a nest of leaves or grasses.

Breeding occurs in midsummer; implantation of a single fertilized egg usually is delayed until November. Development of the young requires 120 days. Four young are born, which are identical quadruplets and all of the same sex; this results from a single egg being fertilized by a single sperm cell and dividing into 4 separate growth centers, each producing an embryo. This feature of reproduction in some species of armadillos is unique among mammals.

The nine-banded armadillo apparently has the unusual capability of crossing relatively deep and wide streams by swimming and/or walking underwater. It can do this because of its high specific gravity (its heavy armor) and the ability to remain submerged for up to 6 minutes without taking in oxygen.

Armadillos have few natural predators, for they are protected by their armorlike covering and the ability to run and dodge and dig rapidly; automobiles, however, take a heavy toll. Contrary to popular belief, they do not curl into a ball when threatened. Armadillos are susceptible to leprosy, making them valuable in medical research.

Rabbits and Hares
Order Lagomorpha

Popular in folklore, legend, and fairy tales, lagomorphs are among our most familiar and readily recognizable mammals. The order contains 2 families: Family Leporidae which includes rabbits and hares, and Family Ochotonidae, the pika. In North America, pikas occur only in the mountains of the western United States; rabbits and hares occur throughout most of the continent. Four species of rabbits, in the genus *Sylvilagus*, and a single species of hare, in the genus *Lepus*, are native to the Carolinas, Virginia, and Maryland. Other species of hares have been introduced into the region at various times in the past, but only the black-tailed jack rabbit has become established.

Rabbits and hares superficially resemble rodents and once were included in the same order. Lagomorphs possess 2 pairs of upper incisor teeth for gnawing. The first pair is rodentlike, each with a distinct groove on the front surface; the second pair, however, is small, peglike, located directly behind the first, and lacks the chiseled cutting edge. The function of the nearly circular second pair of incisors is not clear. As in rodents, there are no canine teeth and a wide diastema is present between the incisors and cheek teeth; the cheek teeth are broad grinding surfaces and are ever-growing, correlated to the animal's herbivorous diet. Other characteristic features of these animals are elongate ears, a rudimentary tail, hind legs that

are longer than forelegs, and hind feet that are significantly larger than the front feet. The ears, hind legs, and hind feet of hares usually are larger than those of rabbits.

Both rabbits and hares produce relatively large litters and are prodigious reproducers; relatively large populations usually are maintained. A rabbit usually builds an elaborate nest, produces altricial young (born blind, scantily haired, and helpless), and cares for its young after birth. A hare utilizes a shallow depression in the soil as a nest, produces precocial young (born with sight, fully haired, and capable of moving about), and provides little parental care.

Large ears and eyes enable rabbits and hares to detect enemies at relatively great distances. Hares tend to have large home ranges within which they feed, and depend upon their keen sense of hearing and speed for escaping danger. Rabbits tend to range over a smaller area, so long as sufficient food, cover, and nesting sites are available.

In spite of their speed and skill in evading enemies, rabbits and hares are important prey for many carnivores and are taken in large numbers. Thus, they are ecologically valuable organisms in the complex food webs that characterize many ecosystems.

Marsh Rabbit
Sylvilagus palustris

Description. The marsh rabbit is usually a darker brown than the more familiar eastern cottontail. It lacks the white "cotton tail" of that species, the underside of its tail being bluish gray instead. The ears of the marsh rabbit are shorter than those of the cottontail, and its small, slender hind feet have long conspicuous claws. The hair on the hind feet is less fluffy than that of the cottontail. The central part of the belly is white, but the rest of the belly is pale grayish brown. The marsh rabbit ranges from 15¾ to 17¾ inches (40 to 45 cm) in total length and weighs 2⅝ to 4⅞ pounds (1.2 to 2.2 kg); the hind foot measures 3½ to 3⅞ inches (9 to 10 cm) in length.

Distribution and Abundance. This rabbit is found from southeastern Virginia southward through the coastal plain of North and South Carolina. There is a single report from the Eastern Shore of Virginia at Hog Island; most larger barrier islands in the Carolinas are inhabited as well. It is common in marshes and bottomlands throughout much of its range, but is localized and relatively uncommon in Virginia and west of the fall line in the Carolinas.

Habitat. The marsh rabbit is aptly named as it usually is found in marshes and swamps. It is often abundant in the brackish marshes of the coastal zone and may be found in wooded swamps and floodplains throughout the coastal plain. It is an excellent swimmer and readily takes to water. Marsh rabbits are often found on isolated islands in coastal river mouths and sounds.

Natural History. Marsh rabbits are generally nocturnal, feeding on a variety of marsh plants such as catbrier, centella, marsh pennywort, cattails, rushes, and cane. They also consume the leaves and twigs of woody plants.

These rabbits are thought to breed throughout the year, and may raise several litters each year. Females give birth to 3 to 5 young after a gestation period estimated to be between 30 and 37 days. The young are born in a depression lined with grasses and fur. They are blind and helpless at birth, but have a well-developed coat of fur. They do not leave the nest until after they are weaned. Many females will breed at less than 1 year of age.

Marsh rabbits are preyed upon by many predators such as great-horned owls, marsh hawks, gray foxes, alligators, and snakes. They are also taken by rabbit hunters who hunt the brackish marshes and coastal plain lowland forests. The high reproductive poten-

Marsh rabbit (Sylvilagus palustris).

tial of these animals appears to make the species quite capable of maintaining high populations in spite of these heavy losses. A much more significant danger appears to be the loss of marsh and swamp habitat.

Eastern Cottontail

Sylvilagus floridanus

Description. This is the best-known rabbit of eastern North America. The upper body is generally reddish brown and the underparts are white; specimens from Smith Island and Fishermans Island, Virginia, are pale sandy-brown in color and those from the western panhandle of Virginia are more grayish. The underside of the short fluffy tail is also white—hence the common name. There is a distinct rusty nape patch and often a white spot on the forehead which help to separate this species from some of the other rabbits in the region. This rabbit usually weighs 2 to 4 pounds (0.9 to 1.8 kg) and has a total length of 12⅝ to 19⅝ inches (32 to 50 cm); the hind foot is 3⅛ to 4⅜ inches (8 to 11 cm) long.

Distribution and Abundance. This is the most widely distributed member of the genus *Sylvilagus* in North America. It is common throughout this region from the barrier beaches to the mountains. Populations fluctuate widely from place to place and from one year to the next.

Habitat. Eastern cottontails are primarily animals of disturbed environments, preferring old fields, brushy

Eastern cottontail (Sylvilagus floridanus).

edges, and other habitats characterized by mixtures of herbaceous and shrubby plants. The availability of dense patches of escape cover apparently is important. Within these habitats, cottontails may be found on coastal islands, in old fields growing up to weeds, and in a variety of other successional and transitional habitats throughout the region. They regularly inhabit the suburbs of most towns and cities and are apt to be encountered almost anywhere.

Natural History. This rabbit eats a variety of foods. During spring and summer herbaceous plants such as Kentucky bluegrass and clover are usually selected, but garden vegetables and field crops (especially soybeans) may also be eaten. In winter these rabbits may eat woody plants, such as staghorn sumac, red maple, apple, and blackberry, although herbaceous perennials continue to provide the bulk of the food in most areas.

The eastern cottontail is well known for its prolific reproductive capability. In the Deep South they may breed all year, but in the Carolinas, Virginia, and Maryland the breeding

Eastern cottontail (Sylvilagus floridanus), *juveniles in nest.*

season probably extends from late winter until fall. The gestation period is about 28 days, and the litter size is usually from 3 to 5 although litters of up to 10 have been reported. A few females may breed during their first year, but most begin producing young during their second summer. Some females have been reported to have as many as 7 litters in a year.

Young eastern cottontails are covered with a coat of fine hair at birth. Their eyes remain closed until they are about a week old, and they do not leave the nest until they are about 2 weeks old. Females construct nests in holes in the ground. Nests are usually built of local plant material and are lined with fur plucked from the mother's belly.

Cottontails are an important food source for such predators as foxes, bobcats, several species of hawks and owls, and the larger snakes; they are taken regularly by domestic cats and dogs. Man is also an important predator, for the eastern cottontail is the most widely hunted game mammal in the eastern United States. In spite of having many enemies, cottontails are able to maintain their populations if adequate suitable habitat is available.

Rabbits contract several diseases including tularemia, a disease that may be transmitted to people coming into contact with the flesh or blood of an infected animal. Ticks and fleas that infest rabbits sometimes carry Rocky Mountain spotted fever, a disease which is becoming more abundant in the region and is passed to humans by the bite of these parasites.

New England Cottontail
Sylvilagus transitionalis

Description. The New England cottontail is very similar to the eastern cottontail in size and color, but is slightly smaller and often has a small distinct black patch of fur (never a white spot) between the relatively short ears. The upperparts are pinkish buff with an overlying black wash, and the ears usually are edged with black. The belly and underside of the tail are white, and the nape is pale reddish brown. The New England cottontail is about 15⅜ to 16⅛ inches (39 to 41 cm) in total length and weighs about 2¼ to 3⅛ pounds (1 to 1.4 kg).

Distribution and Abundance. In the southern part of its range, the New England cottontail occurs only in the mountains. Records for this four-state region are few, but the species has been reported from western Maryland, Virginia, and North Carolina, where it usually occurs in heavily wooded habitats and mountain balds above 2,500 feet (762 m) in elevation.

New England cottontail (Sylvilagus transitionalis), *juvenile.*

There are no records of this species from South Carolina, but a record from northeastern Georgia suggests that it may occur in the mountain counties of South Carolina.

Habitat. The New England cottontail is an animal of higher elevations in the Appalachian Mountains. Key habitat features appear to be the presence of a thick cover of mountain laurel, rhododendron, or blueberries, and coniferous forests. They have been reported to reach highest densities in scrubby habitat 5 to 10 years after clearcutting.

Natural History. Foods eaten are apparently similar to those consumed by the eastern cottontail, although there is some indication that the diet of the New England cottontail is less varied.

New England cottontails breed during the spring and summer, and 3 to 4 litters are produced each year. The gestation period is about 28 days and litters average 3 to 5 young.

There is concern that this species is declining in numbers over portions of its range. Although it was once continuously distributed in the Appalachian Mountains, the New England cottontail is now restricted to isolated areas of suitable habitat where competition with the eastern cottontail is minimal. The New England cottontail, however, remains one of the least-known rabbits, due in part to the difficulty of separating it in the field from the more ubiquitous eastern cottontail.

Swamp Rabbit
Sylvilagus aquaticus

Description. The swamp rabbit is the largest member of the genus *Sylvilagus*; it weighs up to 6 pounds (2.7 kg) and is almost 3 inches (7 mm) longer in total length than the more familiar eastern cottontail. The upper surface of this rabbit is usually a dark rusty brown and the underparts are white. Its tail is slender and somewhat more thinly haired than that of other cottontails. There is a black spot between the ears and never a white spot on the forehead. Total length is about 18 to 22 inches (45 to 55 cm), and these large rabbits weigh 4¼ to 6 pounds (1.9 to 2.7 kg).

Distribution and Abundance. This is an animal of the Deep South and has been reported in this region only from western South Carolina in Oconee,

Pickens, and Anderson counties and from Clay County in southwestern North Carolina. It is likely, however, that it is more widespread in South Carolina and perhaps in southwestern North Carolina than published records indicate. Careful studies and observations in the field, including reports by rabbit hunters, may extend our knowledge of this species in the region.

Habitat. The swamp rabbit generally occupies floodplains, river swamp forests, and canebrakes and is seldom found far from water. These rabbits are good swimmers and take readily to water, as does the marsh rabbit.

Natural History. Although these rabbits feed on a variety of food items, including many sedges and grasses, the use of cane for food has given it a widely accepted local name of "cane-cutter."

Swamp rabbit (Sylvilagus aquaticus). *Photograph by William E. Sanderson.*

The breeding biology of the swamp rabbit is similar to that of other rabbits. The breeding season is long, apparently extending throughout most of the year in the Deep South but being restricted to the spring in more northern parts of its range. A nest is built on the surface of the ground rather than in a burrow, and after a gestation period of about 37 days, 1 to 6 altricial young are born. The distribution and biology of this species is not well known in this region, and additional study is needed. A primary danger facing this species in the Deep South is the drainage and conversion of swamp lands to other uses, eliminating habitat for this species.

Snowshoe Hare

Lepus americanus

Description. This is the only hare native to the eastern United States. Identification is easy in winter as the pelage is completely white except for the tips of the ears, which remain dark. In summer the upperparts are rusty brown, and the tips of the ears are black. Underparts are white to grayish; the tail is white on the upper surface and gray below. The hind feet and ears are large, and, especially in winter, the soles of the feet are well furred. The snowshoe hare averages 3⅛ to 4 pounds (1.4 to 1.8 kg) and 14 to 20 inches (36 to 52 cm) total length; the hind foot is 4⅜ to 5⅞ inches (11 to 15 cm) long.

Distribution and Abundance. Historically, the snowshoe hare occurred in high elevation forests in the Appalachian Mountains of Maryland and Virginia. Individuals taken in Canada have been released in both of these states in an effort to maintain viable populations of this species, though apparently with little success in western Maryland, where it has not been taken in almost 30 years.

Habitat. Little is known about its habitat preferences in this region, but the snowshoe hare appears to prefer the spruce-fir forests and rhododendron thickets of the high mountains. Farther north, where it is more abundant, it occupies a variety of habitats, but is usually most common in early successional stages of dense second growth forests.

Natural History. Succulent herbaceous plants form the major portion of the summer diet, whereas woody plants become more important food items in winter.

The breeding season usually begins in late winter and may extend into late summer. The gestation period is

Snowshoe hare (Lepus americanus), *adult in summer pelage.*

about 37 days. Litter size varies considerably but usually ranges from 2 to 4 with as many as 9 occasionally born. Females may produce 2 to 4 litters per year. The young are born fully furred and with open eyes; they are weaned at about 4 weeks of age.

The snowshoe hare is preyed upon

by many predators, such as bobcats, foxes, hawks, and owls, and it is considered an important game species in New England, Canada, and Alaska, where it is relatively abundant. These hares exhibit a 10-year cycle of abundance in more northern portions of their range, becoming numerous, then scarce, then numerous again over approximately 10-year periods. Little is known of their ecology in the southern Appalachians, but they probably share the problems of other rabbits in mountain habitats. Their greatest threat in this region may come from modification of the few units of suitable habitat available.

Black-tailed Jack Rabbit
Lepus californicus
(Introduced)

Description. This large hare is easily separated from the native rabbits by its large ears and hind limbs. The black-tailed jack rabbit has a black stripe extending from the dorsal surface of the tail onto the back and black edging on the ears. The long ears are 3⅞ to 5⅛ inches (10 to 13 cm) in length, and the hind feet are 4¼ to 5⅞ inches (11 to 15 cm) long. The total length of these animals is from 18½ to 24¾ inches (47 to 63 cm). Adults may weigh 5⅛ to 5½ pounds (2.3 to 2.5 kg) with males being smaller than females.

Distribution and Abundance. The black-tailed jack rabbit is native to western North America, but has been introduced into many eastern states in-cluding the Carolinas, Virginia, and Maryland. The species is locally common in the vicinity of Cobb Island in Northampton County, Virginia, and may be established on the Delmarva Peninsula, for in 1977 a black-tailed jack rabbit was killed by a hunter in Kent County, Maryland.

Black-tailed jack rabbit (Lepus californicus).

Habitat. This is an animal of arid open grasslands in the west. It occurs on coastal barrier islands in Virginia in habitats somewhat similar in aspect and has been taken in Maryland in nurseries of ornamental juniper.

Natural History. This jack rabbit feeds on a wide variety of succulent grasses and forbs during spring and summer, changing to a greater amount of woody plants in the diet in fall and winter. Little is known about its reproductive biology in this region, but in the west the gestation period is 40 to 47 days, and a litter usually comprises 2 to 4 fully furred and precocial young; 2 to 7 litters are produced each year.

A game animal or pest depending on local circumstances, the black-tailed jack rabbit is hunted in many western states where it frequently competes with livestock for food on rangelands. It may carry tularemia (rabbit fever). Such introductions of exotic animals are often a mixed blessing, but most often the negative results far outweigh the positive ones.

Gnawing Mammals
Order Rodentia

This is the largest and most diverse order of mammals, containing approximately 40 percent of all living species. The order is cosmopolitan in distribution and occurs throughout the Carolinas, Virginia, and Maryland, from sea level to the mountaintops. Rodents range in size from small mice, such as the harvest mouse, weighing only ¼ ounce, to the capybara of South America, which weighs up to 110 pounds (50 kg); the largest in the region is the beaver. The order is represented in this region by a total of 30 species in 22 genera and 7 families.

Rodents are most readily distinguishable from other mammals by the structure of their teeth. They must gnaw to live, mostly on vegetation such as grass, bark, and seeds, though many species also include animal matter, mostly insects, in their diet. They have upper and lower pairs of incisor teeth that are arc-shaped and chisel-edged, adapted to cutting and gnawing food. Each incisor is rootless and grows continuously; its anterior surface is covered with hard enamel whereas the remainder is soft dentine. Gnawing wears away the backs of these teeth more rapidly than the fronts, producing a sharp front edge and chisel shape. Without constant wear the incisors would become too long and prevent the animal from feeding. Rodent jaws have no canine teeth, giving rise to a wide space, or diastema, between the incisors and cheek teeth of each jaw. The cheek teeth are used for grinding; in rodents such as voles that chew highly abrasive materials, these teeth also are ever-growing, compensating for wear. In other rodents, the cheek teeth are rooted and become fixed in the jaw.

Rodents have adapted to most terrestrial habitats, and structural variations, especially of the appendages, correspond to their diverse habits and modes of locomotion. Most species are nocturnal, but others, such as some squirrels and chipmunks, are completely diurnal. Many rodents, such as the chipmunk, are hibernators. For most, dormancy is brief; for a few, such as the woodchuck, it is long and deep.

The rodents distributed in the Carolinas, Virginia, and Maryland are divided among 7 families.

Family Sciuridae includes the squirrels, which are rather generalized rodents with long tails and unspecialized bodies. There are 3 groups: tree squirrels and flying squirrels, adapted to arboreal life, and ground squirrels (the woodchuck and chipmunk), adapted to living in burrows.

Family Castoridae includes a single species, the beaver, famous for its aquatic adaptations and engineering skills as a builder of dams and lodges in streams or other bodies of water.

Family Cricetidae is the largest of all families of mammals. In the four-state region it includes 15 rather diverse species in 2 broad groups (or subfamilies): mice and rats with long tails, large eyes and ears, and pointed snouts; and those usually with short tails, small eyes and ears, and blunt snouts.

Family Muridae includes, in the Carolinas, Virginia, and Maryland, the black and Norway rats and the house mouse. These rodents originated in Europe, Asia, and Africa, but as "human commensals" they now are distributed widely on all continents. They are both companion and foe to humans. Although they usually live in the wild, they sometimes share living space with humans in dwellings, farm buildings, warehouses, and the like, often damaging property; they significantly damage crops, despoil stored supplies of grains and other foods, and contaminate water and soil. Additionally, these rats and mice, or the parasites they harbor, transmit a variety of communicable diseases to humans, including bubonic plague, typhus, rat-bite fever, scarlet fever, and others.

Murids differ from the cricetid rodents, which they resemble, in having relatively smaller eyes, a more pointed nose, and a naked tail; the crowns of their cheek teeth bear 3 longitudinal rows of cusps rather than 2, as in many cricetids.

Family Zapodidae includes 2 species of jumping mice in this region, which are distinguished by their long hind legs and tails, and unique and swift mode of locomotion.

Family Erethizontidae is represented in the United States and Canada by a single species, the porcupine. This improbable animal is easily identified by its imposing arsenal of modified hairs called quills. It has been extirpated from the region.

Family Myocastoridae is represented in the region by a single introduced species, the nutria, from South America. It is a large semiaquatic species, resembling a beaver with a long, scaly, ratlike tail.

Eastern Chipmunk
Tamias striatus

Description. The background coloration of the upperparts of the body is reddish brown, becoming rust on the rump and flanks. There are 5 dark brown longitudinal stripes, 1 that is middorsal, extending from the back of the head to the rump, and 2 on each side from shoulder to rump, separated by a yellowish white band. On the face, a single dark brown stripe runs longitudinally through each eye and is bordered above and below by white or buff stripes. A reddish brown stripe is on each cheek. On the sides and feet the color grades to buff and tan; the belly is white. Chipmunks from western Maryland are paler in color, and those from the Carolinas brighter, than chipmunks from elsewhere in the region. The short ears are rounded and erect. The tail is flat, well haired but not bushy, brownish black above and rusty below, with a narrow gray or yellowish fringe. Large, paired cheek pouches open into the mouth. Total

Eastern chipmunk (Tamias striatus).

length ranges from 8⅝ to 10¼ inches (22 to 26 cm), including a tail that is from 3 to 3⅞ inches (7.5 to 10 cm). Weight of adults ranges from about 2⅞ to 4⅜ ounces (80 to 125 gm).

Distribution and Abundance. The eastern chipmunk occurs throughout Maryland and Virginia. In North Carolina its range extends from the northeastern coastal plain and central piedmont to the western border of the state, and in South Carolina its range includes only the western segment of the state. The species is abundant within its range where there is favorable habitat and adequate food.

Habitat. This active ground squirrel inhabits deciduous woodlands, the edges of forests, or open and brushy forests where there are abundant crevices for refuge and suitable structures to conceal burrows, such as rock piles or ridges, brush or log heaps, or roots and stumps.

Natural History. Individuals usually excavate a burrow system within which they live. The main burrow entrance opens into a tunnel which leads to a chamber approximately 3 feet (0.9 m) below ground surface and contains a nest of leaves and stored food. Two or more side tunnels and entrances may be present, but usually are

plugged with dirt and rarely used. The main entrance usually is well hidden and does not have a mound of dirt around its lip as do those of woodchucks and some other burrowing rodents. Most eastern chipmunks tend to be sedentary and may occupy the same burrow system for several years, perhaps for life. These animals generally are not social and even though home ranges may overlap to some degree, individuals maintain exclusive territories around their burrows. Burrows are not shared, except briefly by mother and young.

Eastern chipmunks are diurnal and the intensity of their activity tends to be greatest in early morning and late afternoon, when they focus their activity on food gathering and storage. The chief items of diet are seeds, nuts, acorns, and berries; they also may eat such animals as insects, small amphibians, and birds. They usually forage for food on the ground but may climb trees on occasion. Most food caches are hoarded in great quantity within the burrow system, but some reserves also are scattered above ground. Various types of food are transported to caches in cheek pouches that often bulge prodigiously—each has a capacity equal to a heaping tablespoon.

Activity declines with the advent of lower temperatures in the fall, and the animals hibernate underground from late fall to early spring. Torpor may be deep, with lowered body temperature, breathing rate, and heart rate, for periods of several days, followed by periods of arousal during which an animal may feed and even emerge briefly from the burrow. Torpor is more pro-

nounced in northern parts of their range; in the south activity may continue year round.

Breeding commences when the animals emerge from their burrows in the spring. Most matings are from late February to early April and from late June to early July, producing 2 litters a year. Litters of 4 or 5 blind, hairless young are born after a gestation period of 31 to 32 days. Their eyes are open and the body is well furred after a month; they emerge above ground after 5 to 7 weeks and within 2 weeks establish their own burrow nearby. The young achieve adult size by 3 months, but usually are not sexually mature until after their first winter.

Eastern chipmunks are preyed upon by a variety of predators, including weasels, foxes, bobcats, domestic cats, certain hawks and owls, and numerous snakes, and they are an important component in the food chains of many wildlife communities.

Woodchuck
Marmota monax

Description. The woodchuck or "groundhog" is the largest member of the squirrel family in the region. Its total length ranges from 16½ to 26⅜ inches (42 to 67 cm), including a relatively short tail of only 3⅞ to 6¼ inches (10 to 16 cm). Adults weigh from 5⅛ to 11⅞ pounds (2.3 to 5.4 kg). Females are somewhat smaller and weigh less than males. Woodchucks are heaviest in the fall before hibernation due to stored fat; upon emergence from hibernation in the

Woodchuck (Marmota monax).

spring they will have lost from a third to a half of that weight.

The body of a woodchuck is broad and somewhat flat and squat, with short, powerful legs, and a bushy, flattened tail. The blunt nose, medium-sized eyes, and small rounded ears are set high on the broad head of the animal. Body fur is relatively dense; hairs are long and coarse. The general coloration is grizzled brown or grayish brown. Hairs of the underfur are blackish brown at the bases with tips of pale gray or buff, and the longer guard hairs are strongly tipped with buff white. The top of the head, tail, and feet are darker brown; the belly is lighter, tending to reddish brown, and more sparsely furred than the rest of the body.

Distribution and Abundance. Woodchucks occur throughout Maryland and Virginia, in North Carolina from the mountains in the west to the central piedmont and northeastern tier of counties of the coastal plain, and in South Carolina in the mountainous western tip of the state.

Unlike many species of mammals, woodchucks have extended their range and increased in abundance in response to alteration of the natural environment by humans. They probably were scarce when European settlers first began to clear land, producing open fields, meadows, and fencerows. As habitats suitable to the animals increased, they prospered, as evidenced by their recent range extension onto the Delmarva Peninsula and the piedmont and northern coastal plain of North Carolina.

Habitat. This animal prefers to dig its burrow on the edge of forests that border open land, along brushy fencerows or stream banks, or in grassy fields and meadows. It may also burrow around or under old buildings. The edges of roadways and utility rights-of-way provide additional habitat and probably aided the animal in extending its range.

Natural History. Woodchucks frequently are seen ambling along road sides, foraging for food near the burrow entrance, or sitting on their haunches surveying their domain over the tops of surrounding vegetation. Their burrow is conspicuous because of a mound of excavated dirt around the entrance. The opening may be beneath a rock or stump, in a natural crevice, or completely exposed. Several nest chambers, lined with leaves and grasses, are within the tunnel system. As many as 5 side entrances may be present; these are smaller and usually well concealed, lack dirt mounds, and probably are for emergency use. A single individual or small group, probably a family unit, utilizes a burrow for sleep, rearing young, hibernation, and as a refuge from enemies. Separate burrows may be dug for hibernation and for summer occupancy.

Adults are solitary, but share their home ranges with other woodchucks. The animals are territorial in defending the area around the burrow and use scent and distinctive whistling sounds to warn away intruders. They are most active in the early morning and late afternoon, tending to remain underground during the midday heat of summer.

Woodchucks are vegetarians, rarely eating animal matter. They feed actively from spring to fall and deposit fat up to a half-inch thick over much of the body, especially the back and shoulders. Activity is reduced as cool weather approaches, and by late October to mid-November most woodchucks in mountain habitats have entered their nest chambers to hibernate. Animals from the coastal plain may remain active year round. Those that hibernate pass the winter in a state of deep torpor. Emergence from hibernation occurs in late February or early March and is gradual, with activity above ground increasing as temperatures rise.

The animals mate shortly after emerging from hibernation. Four or 5 blind and naked young, each weighing about 1 ounce (28 gm), are born after a gestation period of about 32 days. Their eyes open after about 4 weeks, and they venture from the nest and burrow after 6 or 7 weeks. The young are cared for by the mother, but after about 2 months she drives them away to establish their own burrows nearby. Young animals are capable of breeding at 2 years of age.

Woodchucks are prey for many carnivorous mammals and large raptors. As a highly visible, ground-dwelling mammal, they are important to people who enjoy animal watching. Perhaps their greatest fame is the folklore that woodchucks venture from their burrows on February 2 (Groundhog Day), and depending upon whether or not their shadow is visible, winter will be shortened or extended. Since they usually emerge into daylight in late February, this otherwise fascinating animal is rather unreliable as a practitioner of the art of weather forecasting. Woodchucks occasionally become pests near farmlands because they consume crops and their burrows are a hazard to livestock; they also burrow into dams and sometimes cause breaks.

Gray Squirrel
Sciurus carolinensis

Description. The gray squirrel is one of the most familiar and visible mammals living in the Carolinas, Virginia, and Maryland. These squirrels range in weight from 9 to $17\frac{1}{8}$ ounces (256 to 485 gm) and in total length from $16\frac{1}{2}$ to $21\frac{5}{8}$ inches (42 to 55 cm), nearly half of which is a slightly flattened, conspicuously bushy tail. Gray squirrels are tree-dwelling animals and have strong legs and relatively long toes and claws for climbing and large eyes that provide both good lateral sight and binocular vision, an advantage when running and leaping among the limbs and avoiding predators.

The overall gray color results from hairs having distinct bands of brown, black, and white, or a "salt and pep-

Gray squirrel (Sciurus carolinensis). *Photograph by Walter C. Biggs, Jr.*

per" appearance. The middorsal area tends to be slightly brownish, continuing darker on the top of the head and cheeks. The ears are tan to light brown. The flanks behind the front legs and the upper surfaces of the feet also tend to be brown to cinnamon. Underparts of the body are white. The long hairs on the tail have brown and black bands and white tips. Coat color shows considerable variation, including reddish and blackish individuals, as well as rare albinos. Gray squirrels from Maryland and the western parts of Virginia and North Carolina are slightly darker in color and larger in size than those from eastern Virginia and most of the Carolinas. The sexes do not differ in color or size. Gray squirrels differ from fox squirrels in being much smaller and more slender and grayer in color and in having white along the edges of the tail.

Distribution and Abundance. The species is distributed widely in the eastern half of the United States and has been introduced in many municipal parks on the West Coast. It is abundant throughout the Carolinas, Virginia, and Maryland, but becomes rare at high elevations in the western mountains or in open oak-pine habitats.

Prodigious numbers of gray squirrels populated the unbroken forests of eastern North America when European colonization began. They destroyed crops, and in places a bounty was paid for their hides. The numbers of squirrels had declined significantly by the 1860s and by the early 1900s there was concern for their survival; clearing, burning, and logging of forests were prime factors in this decline. Fortunately, the reestablishment of closed forests and management efforts have resulted in restoring stable and

abundant populations. This animal currently is one of our most important game species, with approximately 40 million harvested annually in the United States.

Habitat. The preferred habitat of gray squirrels is extensive tracts of mature forests of oaks, hickories, and beeches mixed with other hardwoods and various species of conifers. They frequently are seen along river bluffs and in wooded river bottoms where underbrush is heavy and tree cavities suitable for den sites are abundant. However, they are capable of accommodation with humans and are common in parks and residential areas where adequate food and shelter are available.

Natural History. Gray squirrels are adept climbers and nest in trees. Nests of 2 types are used, either a tree cavity lined with shredded bark and other plant material or a mass of leaves or twigs located in the fork of a tree or in its outer branches, high above the ground. The latter is used when a cavity is not available or for temporary summer shelter.

These squirrels are active during daylight hours, especially soon after sunrise and in late afternoon. They are active year round, with a peak period during September and early October when food is cached for the winter. Activity declines in very cold and inclement weather. A thickened pelage and a layer of fat provide insulation against low winter temperatures.

Their diet varies with the seasons; they eat mostly plant material, but occasionally take insects and bird eggs. In spring they feed on buds, twigs, and flowers of various trees, and may even drink sap. Fruits, seeds, berries, and mushrooms are added in summer. In late summer, fall, and early winter, abundant food usually is derived from the nut crop of oak, hickory, walnut, and other trees; pine seeds, corn, and such fall fruits as grapes and wild cherries also are important. Nuts that fall to the ground, or are cut and either dropped or carried to the ground, are usually buried individually in shallow holes, then carefully concealed for consumption later. Such caches are the main food supply in winter. The size of local squirrel populations is related directly to the quantity of acorns, nuts, and other foods produced in a growing season, for a woodland must produce about 100 pounds (45 kg) of food per year for each squirrel it supports.

Gray squirrels normally produce 2 litters each year, with mating taking place in midwinter and again in June. Two to 5 blind, naked babies are born after a gestation period of 44 to 45 days. After 5 weeks the young have hair, their eyes and ears are open, and incisor teeth are visible. They are weaned at about 8 weeks of age and are fully self-supporting by 12 weeks. If the mother has a second litter, she will leave the first and move to a new nest. The young of the second litter frequently stay through the winter in the nest with the mother. Females are aggressively protective of their young while in the nest. Sexual maturity is reached at about 10 to 11 months of age, and females produce their first litter when they are 1 year old.

Gray squirrels are preyed upon by certain snakes, birds of prey, and mammals such as weasels, raccoons, foxes, and bobcats; large numbers are taken by hunters. In the late summer or early fall squirrels are often seen with one or more swellings, or warbles, beneath the skin. These are the developing larvae of the parasitic botfly, which emerge by late fall. While unsightly, the effect on the squirrel usually is minimal. Natural mortality and loss of habitat seem to be the principal factors affecting population size. Gray squirrels sometimes are considered pests near human habitations as a result of their feeding and gnawing behavior; more often, however, they are valued by sportsmen and those who enjoy watching wild animals.

Fox Squirrel
Sciurus niger

Description. The fox squirrel is the largest of the North American tree squirrels, typically being 20 to 26 inches (50 to 66 cm) in total length, including a tail of 8⅞ to 14⅝ inches (20 to 37 cm), and weighing between about 1½ and 2½ pounds (737 to 1,200 gm). Individuals from the Carolinas are larger than those from Virginia and Maryland.

Coloration is highly variable in fox squirrels. In the Carolinas they tend to be pale gray, usually with some black on the head and feet; the nose, tips of the ears, and the underside are white. The Delmarva fox squirrel, a race found on the Eastern Shore of Maryland and Virginia, usually is bluish gray without the black face common to fox squirrels further south. In Maryland west of the Chesapeake Bay and in the piedmont and mountains of Virginia and North Carolina, a third race occurs which is characterized by a strong infusion of yellow-brown dorsally; its ears, feet, and the underside of the tail are rusty in color. Some individuals have a white underside, but many from the mountains, particularly in western Virginia, have a rufous-colored belly. The tail of fox squirrels is long and bushy and colored similarly to the upper body. Melanistic fox squirrels may show increased amounts of black along with gray, or be almost completely black (except for a white nose and toes), especially animals from the Carolinas.

Distribution and Abundance. Historically, the fox squirrel occurred throughout the mid-Atlantic region, but its present distribution is more restricted. Local concentrations now are found in the sandhills and coastal

Fox squirrel (Sciurus niger).

plain of South Carolina, the south-eastern coastal plain and sandhills of North Carolina, and the mountains of Virginia; it occurs in more isolated areas in the mountains of North Carolina, northern Virginia, and Maryland. The Delmarva fox squirrel,

considered Endangered, is restricted to the Eastern Shore of Maryland and Virginia.

The species remains locally common in the sandhills of the Carolinas and the mountains of Virginia. It is considered a game animal in these states, and there presently are open seasons for hunting. The Delmarva fox squirrel, once more widespread and abundant, now exists only in small numbers, primarily on wildlife refuges. The best opportunity to see this race is at the Chincoteague National Wildlife Refuge, Virginia, and Blackwater National Wildlife Refuge, Maryland.

Habitat. This animal occurs primarily in mature longleaf pine–oak forests and along the edges of adjacent swamps in the Carolinas. It spends much of its time on the ground and prefers forests with relatively open understories, though it can be found on some golf courses in the coastal plain and sandhills. Because it often dens in hollow trunks, the presence of mature trees seems to be important to its survival. Those living in mountain habitats and in the piedmont and Western Shore of Maryland inhabit dense deciduous forests. The Delmarva fox squirrel is often found in stands of mixed loblolly pine–hardwood forest with minimal undergrowth.

Natural History. This squirrel is not as agile as the gray squirrel and spends more time on the ground. Although the fox squirrel feeds on a variety of mast, fruits, and seeds, it seems to prefer pine seeds, cutting both green and mature cones. Acorns, hickory nuts, and the buds and berries of many plants are included in the diet, as are fungi and insects. Individuals sometimes raid corn fields and may damage grain crops.

Breeding occurs in midwinter, and a litter of 1 to 6 (average 2 to 4) young are born in February or March after a gestation period of 44 to 45 days. Older females occasionally produce a second litter in July or August if that year's food crop was unusually productive. The young are born blind and hairless. Eyes open after 4 to 5 weeks, and young squirrels are weaned after 8 to 9 weeks. They may remain with the mother for another month before becoming independent.

A primary cause of the decline in fox squirrel numbers appears to be loss of habitat. These animals are closely associated with the mature,

open pine forests of the south and open hardwood forests of the north; these forests are declining steadily, thus diminishing habitat for the species. Researchers believe that the Delmarva fox squirrel suffered primarily from the loss of suitable stands of loblolly pine, and perhaps from competition with gray squirrels in marginal habitats. There now is an active management program designed to assist the Delmarva fox squirrel, and there is increasing concern that fox squirrels in the Carolinas may need assistance as well.

Red Squirrel
Tamiasciurus hudsonicus

Description. Red squirrels (also known as "boomers" or pine squirrels) have an average weight of about half a pound (227 gm) and a total length of 11¾ to 13¾ inches (30 to 35 cm). The body is reddish brown above with a white underside; squirrels from the western mountains are conspicuously darker than those from other portions of their range in the region. The dark back and light belly are separated in summer by a distinct lateral black stripe; this stripe is absent in the winter pelage. A prominent white ring encircles each eye. In winter, short tufts of hair are present on the tips of the ears. The slightly flattened tail, which often is held erect like a flag, is about 4¾ to 5⅛ inches (12 to 13 cm) long, shorter in proportion to the remainder of the body than those of the gray and fox squirrels. The tail appears reddish brown like the back, but

Red squirrel (Tamiasciurus hudsonicus).

the long hairs are yellow-white at their tips, next to a band of black.

Distribution and Abundance. In this four-state region red squirrels range from the Appalachian Mountains of the Carolinas northward and eastward through all but the southernmost counties of piedmont and coastal Virginia and the coastal plain of Maryland. They are abundant in mountainous habitats, becoming uncommon and localized in the piedmont of Maryland and northern Virginia. There have been several recent reports of red squirrels from the piedmont of North Carolina, with isolated populations in the Winston-Salem and Greensboro areas.

Habitat. Within the predominately mountainous habitats of the region, the red squirrel occurs in both coniferous and hardwood forests, or in

mixed conifer-hardwood stands. They are particularly abundant where there are spruce and hemlock trees.

Natural History. Red squirrels are more territorial than other tree squirrels of the region and are highly aggressive in defending their territory,

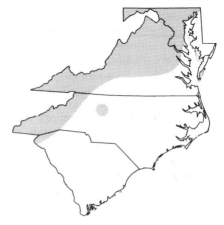

using scolding chatter, threatening postures, and physical attack. Individual territories are sufficiently large to contain several den sites or shelters. A den may be a tree hollow lined with leaves, bark, and fur or a leaf-and-twig nest constructed in branches near the top of a tree, similar to that of the gray squirrel. Red squirrels also occasionally live in burrows under rock piles, stumps, or fallen trees; such burrows contain a nest chamber and connect to caches of food.

Red squirrels are active throughout the day, spending much time scurrying about on the ground and on tree trunks. They continue to move about in winter and even burrow through snow to reach food stores. They eat a variety of foods, but their main diet consists of conifer seeds and hardwood nuts; they also feed on other types of seeds, berries, mushrooms (which they dry on a tree branch before storing), and even insects, snails, and bird's eggs and fledglings. Cones are cached in large quantities in tree hollows, underground, or in open depressions. The vociferous and aggressive behavior of these animals usually prevents loss of their hoard to other squirrels.

These squirrels ordinarily are solitary, but a female allows neighboring males into her territory in late winter when she is ready to mate. After a gestation period of 38 to 42 days, she produces a litter of 1 to 7 (usually 6) naked and blind young. Their eyes open in 4 weeks and they are on their own after about 3 months. Typically, a single litter is produced each year; however, some females may produce a second litter in August or September,

and these offspring stay with the mother over the winter. Individuals breed when 10 to 12 months old.

Although most red squirrels die during the first 2 years after birth, individuals may live 5 years or more. They may be eaten by hawks, owls, bobcats, and other carnivores. These energetic and spirited squirrels are familiar to park and woodland visitors and add fascinating variety to the mammalian fauna of the region.

Southern Flying Squirrel
Glaucomys volans

Description. Flying squirrels are small, unobtrusive creatures and perhaps the most unusually adapted of the squirrel species in the Carolinas, Virginia, and Maryland. They are fully nocturnal, emerging from tree-cavity nests only after their forest habitat has been enveloped in darkness; therefore, they are seen infrequently in the wild and are unfamiliar to many people.

A distinctive structure of flying squirrels is the gliding membrane on each side of the body, formed by a loose fold of well-furred skin that extends from the wrists of the forelimbs to the ankles of the hind limbs. When the legs are fully outstretched, the membranes form a winglike surface that enables the animal to glide downward from tree to tree or from tree to ground. The tail is an effective rudder that helps to control the path of the glide; it is flat and broad, with parallel sides and a rounded tip.

The southern flying squirrel is the smallest of the squirrels in the region.

Southern flying squirrel (Glaucomys volans).

Adults weigh 2⅛ to 3½ ounces (60 to 100 gm) and vary in total length from 8¼ to 9⅞ inches (21 to 25 cm), including a tail of 3½ to 4¾ inches (9 to 12 cm). This squirrel has prominent, rounded ears and a short, somewhat upturned nose. It has dense fur that is soft, fine, and silky in texture. The hairs of the upper body are slate gray, tinged with cinnamon, grading on the sides to a black border at the edges of the gliding membrane. Large, lustrous black eyes are ringed by a narrow band of black. The underparts of the body, including the cheeks, side of the neck, and underside of the legs, are pure to creamy white. The furry tail is uniform gray above and pinkish cinnamon below. The coloration is the same for males and females; both sexes are somewhat darker and browner in summer than in winter. Animals from South Carolina and western North Carolina are darker in color than those from Maryland, Virginia, and central and eastern North Carolina.

Distribution and Abundance. The species is distributed throughout the Carolinas, Virginia, and Maryland with the apparent exception of the coastal barrier islands. This squirrel is relatively common in the region, more so than most people realize.

Habitat. Flying squirrels are active at night in the same type of habitat in which their larger and more visible relatives, gray and fox squirrels, move about during the day. They live in mature hardwood and mixed conifer-hardwood forests, especially where there is an abundance of old trees with natural cavities or woodpecker holes where nests can be built.

Natural History. Southern flying squirrels are most likely to be seen in the soft light of a full moon on a warm summer night. They are active year round, but activity may decline in the severe cold of winter when movement is related to feeding. Cavities in trees with entrances of 1½ to 2 inches (38 to 51 mm), including those made by woodpeckers, are preferred nest sites. In the sandhills area of the Carolinas, flying squirrels and red-cockaded woodpeckers occasionally compete for the same nesting sites. The squirrel may usurp cavities excavated in the trunks of mature pine trees by the bird, an Endangered Species.

Nests are constructed mostly of finely shredded bark and grass, but also of moss, leaves, and feathers. They may be located up to 40 feet (12 m) above the ground, but most often are at heights from 14½ to 20 feet (4.4 to 6 m). Flying squirrels sometimes use outside leaf nests, including those abandoned by gray or fox squirrels; they also may nest in birdhouses, attics, or under the eaves of buildings.

Gliding locomotion is characteristic of flying squirrels. A typical glide may

cover about 20 to 30 feet (6 to 9 m), but glides of 100 feet (30 m) have been observed. The path of the glide is controlled by movements of one or both sides of the membrane and the tail, and turns of 90 degrees or more are possible, enabling the animals to avoid obstacles they encounter. These tiny acrobats land on their hind feet in a head-up position, ready to scamper up a second tree for another glide or to forage on the ground. In addition to gliding, they run or hop on the ground, run along tree limbs, and leap from limb to limb in the manner of other tree squirrels.

Flying squirrels eat acorns, nuts, berries, fruit, seeds and grains, and buds and blossoms in the spring; they may also consume some animal matter, such as insects, birds' eggs and nestlings, and the flesh of dead carcasses. They store nuts for the winter in the nest, in forks and cavities of trees, and on the ground; hoarding behavior is at a peak in November.

Breeding is in January and February and again in June and July; however, not all females produce 2 litters in a year. The gestation period is 40 days; litters commonly include 2 or 3 offspring, but the number may range from 1 to 6. Young are born blind, hairless, with both eyes and ears closed, and pink in color. The gliding membrane is present but transparent. The eyes open after 4 weeks, and by then the body is covered with hair. The young are weaned after 6 to 8 weeks. Mother and young remain together until the next litter is produced. The female alone cares for the youngsters and she is highly protective. Sexual maturity apparently is attained after 1 year.

These are highly social animals. In winter, several individuals, perhaps as many as a dozen or more, congregate in a single nest, an advantage in keeping warm. They are relatively docile and are easily kept as pets.

Flying squirrels are preyed upon by a variety of mammals, tree-climbing snakes, and birds of prey. The greatest threat to the maintenance of significant populations of this interesting mammal is loss of mature trees that provide suitable nesting sites.

Northern Flying Squirrel
Glaucomys sabrinus

Description. The northern flying squirrel closely resembles the southern flying squirrel but is larger in size and more richly colored. In North Carolina the species ranges from about 3⅜ to 5 ounces (95 to 140 gm) in weight and 10¼ to 12¼ inches (26 to 31 cm) in total length. Hairs on the underside of the body are gray at their bases, not pure white as in the southern flying squirrel.

Distribution and Abundance. The northern flying squirrel is distributed widely across northern North America, and a portion of its range extends southward along the peaks of the Appalachian Mountains into western Virginia and North Carolina. The species has a limited distribution in these states and has been found only at isolated localities, specifically, Whitetop Mountain in Virginia and Roan Mountain, Bald Knob near Mount Mitchell, and a few areas in the Great Smoky Mountains in North

Northern flying squirrel (Glaucomys sabrinus). *Photograph by Nancy M. Wells.*

Carolina. These appear to be small, fragmented populations, relics of an earlier period of time when the climate was cooler and this squirrel was more widely distributed in the region. The species has been listed as Endangered in Virginia, in North Carolina it is considered to be Threatened, and the U.S. Fish and Wildlife Service is currently reviewing a petition to classify these southern Appalachian populations as federally Endangered. Its numbers are declining because of long-term vegetational changes as well as alteration of the habitat by humans through such activities as logging and clearing of land. In addition, the southern flying squirrel, whose range overlaps portions of that of the northern flying squirrel and which is more aggressive in behavior, may be displacing its northern relative in some habitats. Another adverse effect of the interaction between these species is that parasites of the southern flying squirrel may be lethal to or severely weaken the northern species, reducing its chances of survival.

Habitat. In the southern Appalachians, the northern flying squirrel is found most often living at altitudes above 5,000 feet (1,525 m) in spruce-fir forests and in mixed forests of conifers and hardwoods.

Natural History. Southern and northern flying squirrels are similar in most aspects of their life histories. The northern flying squirrel seems to have a somewhat more varied diet, relying heavily upon lichens and fungi, but also eating a variety of seeds and nuts (acorns, conifer seeds, beechnuts, and cherry pits), buds, fruit, insects, and animal flesh. It is less dependent on seeds than its southern counterpart.

Reproductive data are sparse, but seem not to be significantly different from that for the southern flying squirrel. Nests in tree cavities or abandoned bird or tree squirrel nests are used, often by small family groups. The animals mate in spring and give birth to 2 to 6 young in May or June. The gestation period is 37 to 42 days.

Beaver

Castor canadensis
(Extirpated and reintroduced)

Description. The beaver is the largest native North American rodent; adults may weigh as much as 60 pounds (27 kg) and are usually 40 to 49 inches (102 to 122 cm) in total length. The beaver is easily recognized by its dark brown color, large webbed hind feet, and broad hairless horizontally flattened tail. It is well adapted for life in the water, having a very dense, soft

Beaver (Castor canadensis).

underfur overlain by longer, coarse guard hairs; a thin nictitating membrane provides a transparent covering for the eyes when the animal is submerged.

Distribution and Abundance. The beaver apparently occurred throughout North America at the time of settlement by Europeans, but was trapped extensively and was extirpated from most eastern states by the beginning of the nineteenth century. It has been widely reintroduced, however, and again inhabits the region. Distribution is sporadic at present, there often being a mosaic of occupied and unoccupied areas. Numbers are increasing, however, and the beaver is continuing to reoccupy its previous range.

Habitat. This semiaquatic mammal usually lives along small wooded streams which it often dams to form shallow impoundments called beaver ponds. Dams are built of sticks, twigs, small logs, and mud. Such dams may be only a few feet in length or many yards long depending primarily upon local topography. Beaver often improve on man-made causeways across

Beaver dam in coastal plain floodplain.

stream valleys by plugging culverts. Within the impounded area they generally construct large dome-shaped lodges. These have one or more underwater entrances and a living area just above the water level. Beaver living along large streams may burrow and construct dens in stream banks rather than build dams and lodges.

Natural History. Each lodge is usually occupied by a small colony of 4 to 10 individuals, consisting of a male and female and their offspring from 2 previous litters. One or more families may occupy a pond.

Beaver are usually active at night and are seldom seen during daylight hours. They are famous for their ability to cut down trees with their chisel-like teeth, and the twigs and soft tissues of the bark of these trees are an important component of their diet. They also eat many kinds of herbaceous plants that grow in or near their ponds. Such herbaceous vegetation is a major summer food, but the bark and twigs of saplings and trees are primary winter foods.

Young beaver kits are born in the lodge in spring after a gestation period of approximately 4 months. A litter usually consists of 3 to 4 young which are born fully furred and with their eyes partly open. They are weaned at about 6 to 8 weeks of age and usually reach sexual maturity by the time they are 2 years old; this species is monogamous and may mate for life. Wild beaver have been known to live to be 20 years old, but it is probable that only a few reach even half that age.

The beaver has few natural enemies in this region, although the young may occasionally be taken by predators.

Beaver within a colony have a complex system of communication, the best known signal being a loud slap of the tail on the surface of the water. This signal is used as a warning and usually results in the beaver diving beneath the pond surface.

The beaver was important in the early exploration and settlement of this country by Europeans. The soft dense fur was a valuable commodity, and beaver trappers were often the first Europeans to move into new wilderness areas. The high demand for pelts led to the extirpation of the beaver from much of its range in North America. Reintroduction and protection have resulted in the beaver recolonizing much of its former range. Reestablishment of beaver populations has had both positive and negative consequences. Beaver ponds provide wetland habitat for fish, furbearers, and waterfowl. Impoundments also trap sediment, slow erosion, and serve as catchment basins during periods of heavy rains, thus providing some flood control. Beaver dams, however, sometimes flood low-lying farm land, roads, and other man-made structures. Nevertheless, most people seem to consider the renewed presence of the beaver as a desirable result of modern wildlife management practices.

Marsh Rice Rat
Oryzomys palustris

Description. This medium-sized rodent resembles the Norway rat but is smaller and has a tail that is sparsely furred in contrast to the essentially naked tail of the Norway rat. The marsh rice rat might also be confused with the hispid cotton rat, especially when comparing young animals. The tail of the marsh rice rat, however, is about as long as the head and body and the feet are whitish, whereas the tail of the hispid cotton rat is much shorter than the head and body length and the feet are blackish. The marsh rice rat usually has grayish brown upperparts and is paler below. The fur is relatively short and smooth. Marsh rice rats are 7⅛ to 11⅜ inches (18 to 29 cm) in total length and have a long tail of about 3½ to 5½ inches (9 to 14 cm). Adult weight ranges from 1 to 5⅝ ounces (29 to 159 gm), but usually averages 1⅜ to 2⅛ ounces (40 to 61 gm).

Distribution and Abundance. This semiaquatic rodent occurs throughout South Carolina but apparently is restricted to the coastal plain and lower piedmont of North Carolina and Virginia. In Maryland it occurs only in coastal plain counties bordering the Chesapeake Bay. These marsh-dwelling rats are often abundant, especially in estuarine marshes along the coast.

Marsh rice rat (Oryzomys palustris).

Habitat. As the common name suggests, these rats are animals of marshes and marsh edges. They occupy fresh, brackish, and salt marshes, but will venture into upland grasslands adjacent to marshes and at times are found well removed from wetland habitats. Individuals have also been taken along a mountain stream in South Carolina.

Natural History. Marsh rice rats are generally nocturnal animals that feed on a variety of herbaceous plants and some animal matter. They prefer seeds and vegetation at some times of the year, but eat snails, insects, and other arthropods at other times. In coastal salt marshes they feed regularly on smooth cordgrass and are known to eat fiddler crabs and the eggs and young of small marsh-nesting birds.

The breeding season begins in late winter or early spring and extends into late fall. The gestation period is about 25 days and litter size averages 4 to 5. The young are usually born in a globular nest constructed of grasses and often placed over water; they may confiscate the nests of marsh wrens. Weaning occurs at about 11 days; the young become sexually mature at 50 to 60 days and usually breed during their first summer.

Marsh rice rats are eaten by a number of marsh-dwelling predators, especially marsh hawks and short-eared and barn owls. They probably are also taken by a number of snakes and mammalian predators such as mink. They are excellent swimmers and will dive and swim underwater to escape enemies.

These rodents infrequently come into direct contact with humans because their habitats are seldom entered except by hunters and trappers.

They may be startled from hummocks during high tide by autumn marsh hen hunters.

___

Eastern Harvest Mouse

Reithrodontomys humulis

Description. The eastern harvest mouse resembles a small house mouse. It is brownish gray above and has a faint buff wash along the sides; the underparts are grayish white. The long tail is bicolored, gray above and white below. The ears are rather large. The best way to separate it from the house mouse is to look closely at the front surfaces of the upper incisors; harvest mice have grooves down the middle of the incisors whereas house mice do not. These diminutive mice are from 3⅞ to 6 inches (99 to 152 mm) in total length and have tails between 1¾ and 2⅞ inches (43 to 73 mm) long. They usually weigh between ⅛ and ⅜ ounce (5.8 to 11.7 gm). Eastern harvest mice from Maryland and the lower piedmont of Virginia are slightly smaller in size and are grayish in color as compared with the larger reddish mice from other parts of the four-state region.

Distribution and Abundance. Eastern harvest mice apparently occur throughout the Carolinas and Virginia except in the mountains where they are usually limited to lower elevations. In Maryland they have been found only in the southern piedmont. This mouse is abundant in some areas of the southeastern United States but is generally uncommon in our region.

Habitat. These mice occupy old-field habitats. They are most likely to be

Eastern harvest mouse (Reithrodontomys humulis).

small rodents—primarily environmental contamination, losses to predators and diseases, and competition with other species. This is a mouse seldom seen by humans and usually mistaken for a house mouse when it is discovered.

Oldfield Mouse

Peromyscus polionotus

Description. Members of the genus *Peromyscus* that occur in this region can be distinguished from other rodent species by the relatively small size (total length less than 8½ inches or 215 mm), relatively long tail (more than half the head and body length), large black eyes and dark eye rings, coloration of the body and tail (brownish above and whitish below), and ungrooved upper incisors. The eastern harvest mouse and the golden mouse might be confused with *Peromyscus*, but the harvest mouse is smaller in size and has a tan belly and grooved upper incisors, whereas the golden mouse is distinctly ocherous in coloration.

The oldfield mouse is the most distinctive member of the genus *Peromyscus* in the region and is easily identified by its small size and pale coloration. In total length the oldfield mouse measures 4¾ to 5¼ inches (122 to 134 mm), including a tail of 1⅝ to 2 inches (40 to 52 mm); the hind foot is ½ to ¾ inch (14 to 18 mm) in length. The upper parts are pale brownish gray and the tail is sharply bicolored, being predominantly white except for a thin brown-

found in broomsedge fields, but also live in cultivated grain fields and other open areas dominated by tall grasses. They may be found in wet meadows as well as more typical dry upland fields.

Natural History. The food of the eastern harvest mouse consists of the seeds, fruits, and grasses common to its habitat. Although they eat a variety of food items, they appear to exist primarily on seeds. They are excellent climbers and may forage above ground in the dense grasses and weeds that dominate their habitats.

Eastern harvest mice breed throughout the warmer portions of the year. They build a globular nest either on or just above the ground. After a gestation period of 21 to 22 days a litter of 3 to 5 young is born. They are weaned by their fourth week and may breed by 11 weeks of age.

This species has been little studied, and knowledge about population levels and the factors that determine its numbers is scanty. It probably faces most of the same problems as other

Oldfield mouse (Peromyscus polionotus).

ish stripe that extends down the middle of the top of the tail.

Distribution and Abundance. The oldfield mouse is restricted in distribution to the southeastern United States, and part of that range extends northward into the Carolinas. In

South Carolina it occurs throughout most of the southern and western parts of the state where it is locally common, especially in the sandhills; it is not known from the northeastern coastal counties or barrier islands. The northernmost record is from Rutherford County, North Carolina, barely 5 miles (8 km) from the South Carolina border, and it is believed that this species is expanding its range northward along highway rights-of-way. Oldfield mice from west of the fall line are somewhat darker in color than those from east of the fall line.

Habitat. The name of this animal aptly describes its preferred habitat, for sandy fallow fields are its primary residence. Other border habitats, such as those along roadsides and ditches or next to fields of corn, melon, cotton, peanuts, and other cultivated crops, are also utilized. Clay and rocky soils are avoided.

Natural History. Little is known about the natural history of the oldfield mouse. Its presence is indicated by noticeable mounds of soil that mark the entrance to underground tunnels and nests. Most of the day is spent in a burrow, the entrance to which may be plugged with soil during the day or immediately before a heavy rain. An additional escape tunnel is available should the main entrance become blocked by a predator. At night this secretive little burrower emerges in search of seeds, berries, and insects.

The social behavior of the oldfield mouse would be an interesting topic for study. Females evidently are territorial during the reproductive season, which may extend throughout most of the year. Breeding peaks in the autumn and is reduced during the hot-test part of the summer; at least 2 litters, containing an average of 3 or 4 young, are born each year. Oldfield mice make excellent pets. They are very docile, do not bite, and adapt well in captivity.

Deer Mouse
Peromyscus maniculatus

Description. The deer mouse is brownish on the head and sides, with a slightly darker stripe that extends down the middle of the back; the underside and feet are white. The long tail is distinctly bicolored, dark brown above and white below, and a small tuft of hairs conceals its tip. The large, scantily haired ears are dusky in color, their margins edged with distinctly paler hairs. Summer deer mice are noticeably brighter in color than

Deer mouse (Peromyscus maniculatus).

winter animals, and juveniles are dark gray above and white below.

Two races of the deer mouse occur in the Carolinas, Virginia, and Maryland. One has a tail that is slightly longer than the head and body, and its ears and feet are relatively large; the other has a tail that is distinctly shorter than the head and body, and its ears and feet are relatively small. There also are ecological differences between them; these are discussed below.

The long-tailed race measures 6 to 7⅞ inches (152 to 200 mm) in total length with a tail of 2⅞ to 4 inches (72 to 102 mm), whereas the short-tailed race is 5⅞ to 6 inches (149 to 152 mm) in total length including a tail of 2⅜ to 2½ inches (59 to 63 mm). The hind foot of the former is ¾ to ⅞ inch (18 to 23 mm) long, whereas that of the latter seldom reaches ¾ inch (18 mm).

The deer mouse might be confused with both the white-footed mouse and the cotton mouse, but in this region the geographic ranges of the cotton mouse and deer mouse do not overlap. The deer mouse usually can be separated from the white-footed mouse by the following characteristics. The tail of the deer mouse is distinctly bicolored and well haired, particularly the noticeable tuft of hairs at the tip, and the fur is somewhat duller in coloration. The tail of the white-footed mouse is not distinctly bicolored and is less hairy, and it lacks the obvious tuft of hairs at the tip. Tail length also is useful in separating these mice; it is slightly longer than the head and body in the long-tailed race of deer mouse, slightly less than

the head and body in the white-footed mouse, and much shorter than the head and body in the short-tailed deer mouse. Also, the ears and feet of the short-tailed race of deer mouse are relatively smaller than those of the white-footed mouse.

Distribution and Abundance. The long-tailed deer mouse is a denizen of the cool, damp forests in the mountainous section of western Maryland, Virginia, and North and South Carolina, where it is one of the most abundant small mammals. The short-tailed deer mouse is a recent immigrant into the region and is rare and localized; it has been taken in Prince Georges County in Maryland, a few counties in northern Virginia, and near Harrisonburg, Virginia.

Habitat. Coniferous forests are preferred by the long-tailed race, but mixed evergreen-deciduous forests, hardwood forests, and rhododendron thickets are also inhabited. The short-tailed race, however, lives in grasslands, meadows, and agricultural

fields, hence its ecological separation from the long-tailed form. Although both races occur in Maryland and Virginia, they do not come into contact and interbreed.

Natural History. The long-tailed race nests under moss-covered logs, stumps, and boulders, or in unoccupied tree cavities and abandoned squirrel nests. The short-tailed form nests on the ground under rocks, boards, and fallen logs. Both races reproduce from early March until November, although breeding appears to diminish during the hottest part of the summer; a female can have 3 or 4 litters in a year. Two to 7 young are born naked and blind, but grow rapidly and are weaned in about 3 weeks.

At night the deer mouse forages for mast, berries, and snails (long-tailed form) or grain, berries, and seeds (short-tailed form); insects and small vertebrates are consumed by both on occasion. Vegetable matter is stored in caches, to be eaten during times when food is scarce. Deer mice do not hibernate, and are important year-round prey for skunks, foxes, weasels, hawks, owls, and snakes.

White-footed Mouse
Peromyscus leucopus

Description. Like those of other species of *Peromyscus* that occur in Maryland, Virginia, and the Carolinas, the head and sides of the white-footed mouse are brownish, the middorsal stripe is slightly darker, and the belly and feet are white; the line separating the brown sides and white belly is distinct. The tail of the white-footed mouse is indistinctly bicolored, being brownish

White-footed mouse (Peromyscus leucopus).

above and whitish below. The large, dusky-colored ears are narrowly edged with white. The summer pelage is more reddish than the winter pelage. The dorsal fur of juveniles is slate gray. The white-footed mouse is 6 to 7⅜ inches (152 to 188 mm) in total length, including a tail of 2½ to 3⅝ inches (65 to 92 mm); the hind foot measures ⅝ to ⅞ inch (17 to 22 mm).

In areas where 2 or more species live together, mice of the genus *Peromyscus* are difficult to identify to species. In the Appalachian Mountains the white-footed mouse and deer mouse occur together; characteristics of the tail (described in the preceding account) best separate them. In the lower piedmont of the Carolinas, and the Dismal Swamp and James River regions of Virginia, however, it is the white-footed mouse and the cotton mouse that coexist. Means by which the white-footed mouse and the cotton mouse can be separated are described in the following account.

Distribution and Abundance. This is one of the most widespread and abundant small mammals in the region, being distributed throughout Maryland and Virginia and most of the Carolinas. It has not been taken in the lower coastal plain of South Carolina or southeastern North Carolina and occurs only sporadically on the highest mountains in the western part of the four-state region.

Three races of white-footed mice inhabit the region. A brightly colored form of medium size occurs in South Carolina, the piedmont and coastal plain of North Carolina, and the Dis-

mal Swamp and Eastern Shore regions of Virginia. A relatively large and pale race lives in Maryland, the remainder of Virginia, and the Appalachian Mountains as far south as the Great Smoky Mountains. The third form is restricted to the Virginia Beach area and is characterized by its small size and pale color.

Habitat. Hardwood forests are the choice habitat of the white-footed mouse, but field margins, myrtle thickets, marshes, canebrakes, and brushy fencerows also are inhabited. This mouse seldom occurs in grassy fields, and tends to avoid the cool, damp spruce-fir forests in the mountains and the pocosins and sandhills in the coastal plain of the Carolinas. On the other hand, it is one of the most abundant small mammals in the Dismal Swamp region.

Natural History. The white-footed mouse builds a nest of grass, leaves, shredded bark, and other vegetation in hollow trees on the ground or several feet above the forest floor, under rocks or fallen logs, in abandoned squirrel and bird nests, in stumps, and in houses.

One to 7 (usually 4) young are born after a gestation period of 3 to 4 weeks; they grow quickly and are capable of mating in 2 months. Females apparently breed throughout the year and are capable of giving birth to as many as 4 litters in a year. Other aspects of the natural history of the white-footed mouse are similar to those of the cotton mouse and are discussed in the next account.

Cotton Mouse

Peromyscus gossypinus

Description. This mouse is the largest member of the genus occurring in the region, measuring 6⅞ to 7¾ inches

(175 to 198 mm) in total length. The tail is 2¾ to 3½ inches (70 to 90 mm) long and indistinctly bicolored; the relatively large hind foot is ⅞ to 1 inch (22 to 25 mm) in length. The upperparts are orange-brown to dark brown, the midstripe darker than the flanks, and the belly and feet are dull white. Juveniles are grayish above and whitish below.

The cotton mouse and the white-footed mouse are very similar in appearance, and both occur in the swamps of southeastern Virginia and northeastern North Carolina and the lower piedmont of the Carolinas. In these areas the cotton mouse can be distinguished from the white-footed mouse by its darker pelage, stockier build, and larger hind foot. The habitat preferences of these species are somewhat different also, as discussed below.

Cotton mouse (Peromyscus gossypinus).

Distribution and Abundance. The cotton mouse is distributed in the coastal plain of southeastern Virginia and the coastal plain and sandhills of the Carolinas; it also lives on some of the larger barrier islands along the Outer Banks. In the northern part of its range, it occurs only along such coastal river bottomlands as the James, Chowan, Tar, and Neuse drainage systems; in South Carolina its range extends west of the fall line. The cotton mouse is locally abundant throughout much of its range in this region.

Where the cotton mouse and white-footed mouse exist together, numbers of both are unstable; for example, cotton mice were relatively abundant in the Dismal Swamp in the 1890s and 1930s, but rare to absent in the early 1900s and today.

Habitat. The cotton mouse favors lowland deciduous forests, cane and cypress swamps, thickets, and river floodplains, but also inhabits upland pine-hardwood forests, buildings, and scattered piles of vegetation in clear-cut forests.

Natural History. Reproduction occurs throughout most of the year, but diminishes slightly during the hottest and coldest months. Several litters are born each year. One to 7 (average 4) blind, naked young are born after a gestation period of about 23 days; they grow rapidly and are sexually mature after 70 days. The cotton mouse, like other small rodents which occur in the region, is an important part of many food webs. It seldom lives more than 1 year in nature because of predators such as birds, snakes, and carnivorous mammals.

Nests are constructed of various types of vegetation including cotton, hence its common name. Some nests are built in trees, but most are on the ground under logs or rocks, in stumps and brush piles, or behind the bark of rotting trees. This mouse is primarily terrestrial, foraging for berries, seeds, nuts, insects, and other animal material. When river bottoms are inundated by heavy rains or periodic flooding, the cotton mouse readily enters water or climbs trees.

Golden Mouse
Ochrotomys nuttalli

Description. This attractive mouse has soft, thick pelage that differs in color from that of other mice and rats. Adults are tawny or ocherous on the upper body and ears, and creamy white with an ocherous wash on the underparts and feet. The tail is faintly bicolored. The fur of young animals is

Golden mouse (Ochrotomys nuttalli).

slightly darker than that of adults. Adults have a total length of 5½ to 7½ inches (140 to 190 mm), including a relatively short tail of 2⅝ to 3¾ inches (68 to 95 mm); they weigh from ½ to 1 ounce (13 to 27 gm).

Golden mice in the Appalachian Mountains are duskier reddish brown in color and average larger in overall size than those in the piedmont region of Virginia and the Carolinas; also, mice gradually increase in size from the piedmont through the coastal plain. The coloration of the golden mouse is similar to that of meadow and woodland jumping mice, but those species have a much longer tail than the golden mouse. Golden mice also may be confused with the various species of *Peromyscus*, but they differ from them in that the golden upper and paler lower colors lack a clear line of separation on the sides of the body.

Distribution and Abundance. This species is distributed statewide in both North and South Carolina; in Virginia it occurs from the coast to the mountains in the central and southern portions of the state and perhaps northward along the Appalachian Mountains; it does not occur in Maryland or northern Virginia. It is common throughout much of its range, but tends to be less numerous near the coast except in the vicinity of Dismal Swamp where conditions seem to be optimal.

Habitat. Golden mice are found in a variety of habitats, from moist woodlands and boulder-strewn slopes and ridges of the mountains to low thickets, swampy woodlands, and canebrakes at lower elevations. They usually are associated with areas affording ample protective cover or where there

is an understory of bushes and vines, especially honeysuckle, greenbrier, and grape.

Natural History. This arboreal mouse is adept at moving about among vines and small branches. Nests are constructed in shrubs, trees, vines, and even clumps of Spanish moss; they usually are only several feet above the ground, but sometimes may be as high as 30 feet (9 m). Underground nests are used in winter. A nest is about 5⅛ inches (13 cm) in diameter, with an outer layer of leaves and an inner chamber lined with grasses and shredded bark; sometimes the nests of other animals, especially birds, are remodeled. In addition to globular nests, loosely constructed platforms are also used. These serve as places to rest and feed, as evidenced by food residues inevitably present. The main items of diet are a variety of seeds, nuts, and berries, as well as insects.

Golden mice are nocturnal and active year round. They forage near the nest, both on the ground and in the trees and vines in which they live.

Seeds and other food are collected in cheek pouches, transported to feeding platforms, and consumed.

Reproductive activity may continue throughout the year in warmer southern climates and in colder areas may occur from March to October. The animal has a gestation period of 25 to 30 days, and a female usually produces several litters per season; she often is impregnated soon after giving birth. A litter usually consists of 2 or 3 young, though litter size may vary from 1 to 4. The eyes of the offspring open in about 13 days, weaning begins after 17 to 18 days, and the young become independent after about 4 weeks.

Golden mice are easily captured, especially at the nest, and are extremely docile and do well in captivity. They seem to cause no harm and add significantly to the aesthetic qualities of our natural environment.

Hispid Cotton Rat
Sigmodon hispidus

Description. Hispid cotton rats are medium-sized rodents with grizzled brownish gray upperparts and more uniform pale gray underparts. The tail is relatively short, slightly less than half the length of the head and body, and only sparsely haired. The short ears are partially hidden by the coarse fur. Total length is usually between 9 and 12⅝ inches (23 and 32 cm), with males being slightly larger than females; tail length ranges from 3⅛ to 5½ inches (8 to 14 cm). Adult body weight is quite variable extending

Hispid cotton rat (Sigmodon hispidus).

from only 4 ounces to over 8⅝ ounces (112 to 245 gm). Hispid cotton rats from the coastal plain of the Carolinas are darker than rats from the remainder of those states, and rats from Virginia are smaller than those from the Carolinas. These rats, especially when young, might be confused with marsh rice rats; means by which these species can be distinguished are explained in the account of the latter.

Distribution and Abundance. These rodents are found throughout much of the Carolinas and south-central Virginia. They apparently do not occur at altitudes above 3,000 feet (914 m) and are also absent from some coastal islands. Hispid cotton rats are often abundant in appropriate habitat, and since they are active both during the day and night, they are often encountered by people, especially in rural areas.

Habitat. These are animals of grasslands and weedy fields. Thick pastures, grassy roadsides, and abandoned agricultural fields dominated by broomsedge provide suitable habitat. Hispid cotton rats also occupy field edges where there are tangles of hon-

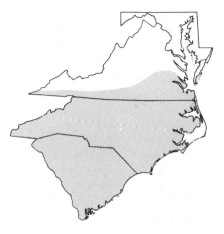

eysuckle and other vines. The key requirement seems to be dense grasses and forbs that provide good overhead cover and an abundant supply of succulent food. A close look at such habitats usually reveals a network of runways beneath the grass canopy. Often such runways are littered with piles of cut grass fragments—a good indication of the presence of these rats.

Natural History. These herbivores feed on a variety of plants, most often eating the leaves and shoots, but also foraging for seeds and excavating the roots and tubers of some herbaceous species. Some insects also are eaten, especially in the fall and winter, but animal matter usually represents only a small portion of the diet. These rodents occasionally girdle small trees, especially when preferred foods are scarce.

Hispid cotton rats are prolific animals. A female normally produces several litters, which average about 5 young each, during each breeding season. They breed from early spring to late fall and may reproduce throughout the year in the warmer parts of their range. The gestation period is about 27 days and the young grow rapidly. They are weaned within 3 weeks and will begin producing young by 2 months of age. They are short lived, however, with most individuals living less than a year.

These mammals provide food for many predators. Their size and abundance make them ideal prey for foxes, bobcats, weasels, hawks, owls, and many snakes. In spite of heavy losses to predators, cotton rats are thought to be one of the most abundant small mammals in the southeastern United States.

Eastern Woodrat
Neotoma floridana

Description. The eastern woodrat has a compact, muscular body that ranges in total length from 13⅜ to 16⅞ inches (34 to 43 cm), including a tail of 6¼ to 7⅞ inches (16 to 20 cm); it weighs 7⅝ to 11¾ ounces (217 to 333 gm). The pelage is rather long and soft. The upper body is colored brownish gray with numerous black-tipped hairs, which produce a grizzled appearance; the fur is darkest at the midline, grading to buff on the sides. The underside and feet are white, there being a distinct line separating the white belly from the darker upper body. The tail is distinctly bicolored, dark brown above and white below. Adults in the summer are more brightly colored above, tending toward cinnamon. Vibrissae are long and conspicuous; the eyes are black, large, and somewhat bulging; and the ears are prominent and sparsely haired.

This rat bears a superficial resemblance to the Norway rat, but the eastern woodrat has larger and more prominent eyes, a blunter snout, less coarse fur, and a more densely haired tail that lacks obvious scales.

Distribution and Abundance. The distribution of the eastern woodrat in the four-state region is disjunct, with populations being most common from western and piedmont Maryland and northern Virginia southward along the Appalachian Mountains to the west-

Eastern woodrat (Neotoma floridana).

ern corner of South Carolina. Speci-mens from the Great Smoky Mountains and western South Carolina are smaller than those from farther north in the region. There also is a pale, medium-sized coastal race whose range extends northward along the coastal plain of eastern South Caro-

lina into extreme southeastern North Carolina at least as far north as Pender County. It is relatively uncommon in this portion of its range.

Habitat. This is a woodland species, preferring deciduous forests. In the mountains it is associated with talus slopes, rocky outcrops, bluffs along river valleys, cliffs with boulders, crevices, or caves. In the coastal plain it may be found in lowland forests, swamps, marshes, and even grass-lands; they often appropriate abandoned buildings.

Natural History. These energetic and resourceful animals build relatively large houses which are their trademark; houses are usually constructed of sticks, twigs, leaves, and a variety of other materials available nearby. A house may be tall and conical in shape if a cliff face or tree trunk provides support, or somewhat flattened in locations where a vertical prop is lacking. Each house may contain 2 or more spherical nests made of grasses, shredded bark, and feathers. Woodrats use their houses throughout the year and keep them clean; fecal droppings and urine are deposited at specific sites away from the nest.

These rodents have an interesting habit of collecting objects that attract their attention, especially shiny and colorful items such as bottle caps, coins, and shotgun shells; hence they sometimes are called "pack rats." They tend to drop whatever they are carrying to pick up something new or more appealing.

Woodrats are solitary except when breeding and rearing young; they generally are intolerant and often chase and fight each other. They are active year round and nocturnal except for occasional activity in late afternoon. Food habits are highly variable; the diet is mainly plant material, depending upon what is available locally. The storage of food, primarily nuts and seeds, is a major activity in September and October; caches are located in galleries at the top of the house.

The breeding season varies geographically; in warmer coastal areas it may be year round, whereas it may be curtailed during the winter months in the mountains. Two to 3 litters may be produced annually; typical litter size averages 2 or 3 individuals, with a range from 1 to 6. The gestation period is from 33 to 35 days. The young are highly altricial; their eyes open in 15 to 21 days and they are weaned after 4 weeks. Young born early in the year may breed the same year, but breeding usually commences the year following birth.

Eastern woodrats have a long life expectancy in comparison to other similarly sized rodents; however, they are preyed upon heavily by owls, skunks, weasels, various snakes, and other predators.

Southern Red-backed Vole
Clethrionomys gapperi

Description. This short-tailed rodent has a total length of 5½ to 6⅛ inches (140 to 155 mm) and weighs from ¾ to 1½ ounces (20 to 42 gm). It has small eyes, prominent ears that extend well above the fur, and a short, distinctly bicolored tail. The length of the tail ranges from 1½ to 2 inches (38 to 50 mm), or less than a third of the animal's total length.

The pelage is rusty red along a broad dorsal area extending from forehead to rump, grading to grayish buff on the sides and face. Underparts are pale gray or silver, a color produced by blackish hairs being tipped with white. The feet are brownish gray. The pelage in summer is usually darker and duskier than in winter, and young animals tend to be darker than adults. Specimens from North Carolina and southern Virginia are slightly larger and much darker in color than those from northern Virginia and Maryland.

Distribution and Abundance. This species generally is restricted to elevations above 2,500 feet (762 m) in the Appalachian Mountains of Maryland, Virginia, and North Carolina. We are unaware of any records in South Carolina, although it may reach into

Southern red-backed vole (Clethrionomys gapperi).

the mountainous tip of that state. This mammal is abundant throughout its range in the region.

Habitat. Southern red-backed voles usually are associated with boreal evergreen forests (especially with spruce and hemlock) having thick ground cover. They most often are found liv-

ing where there are moss-covered rocks, rotting logs, and exposed roots on cool, damp, and shaded slopes. They are occasionally found in grassy clearings, treeless mountain balds, rhododendron thickets, and deciduous forests.

Natural History. This rodent is active year round, even in the coldest part of winter. Although it is mainly nocturnal, it also is often abroad during the day. It does not construct well-defined runways or tunnels of its own, but either utilizes those of moles, shrews, chipmunks, or other small mice or depends on natural cover provided by leaves or grasses. Its food consists of leaves and stems of succulent green plants, nuts, seeds, berries, mosses, fungi, lichens, and ferns. Food is cached in the fall for use during winter.

A nest usually is simple and globular, perhaps 3 to 4 inches (76 to 102 mm) in diameter, and made of such materials as grass stems, leaves, and moss. It may be located in a burrow or natural cavity, or under roots, logs, or a stump; sometimes the nest of some other small mammal may be used.

The breeding season is from early spring to late fall and several litters are produced in succession during the year. The gestation period is 17 to 19 days; a typical litter includes 3 or 4 offspring, though litter size may vary from 2 to 8. The eyes open after 12 to 13 days and the young are weaned at the age of 17 days. Growth is rapid and reproductive maturity is usually reached during the year an animal is born. This vole is often the most numerous small mammal in favorable habitats, and it is an important prey species for many carnivorous mammals, snakes, and birds of prey.

Meadow Vole
Microtus pennsylvanicus

Description. Voles are small rodents characterized by short tails and ear openings that are partly guarded by fur. The meadow vole is the most common vole in eastern North America. It has short dense fur and relatively short ears, but the tail, which is more than twice as long as the hind foot, is longer than that of other species of voles. Color on the back and sides varies from chestnut-brown in summer to dark gray-brown in winter, and the belly fur is dark gray. Immature meadow voles are usually nearly black. Individuals from the lower half of the Eastern Shore, eastern mainland Virginia, and northeastern North Carolina are darker than those from more inland areas. There is also considerable variation in size: individuals from southern Mary-

Meadow vole (Microtus pennsylvanicus).

land, eastern Virginia, and north-eastern North Carolina are somewhat larger than those from other parts of the region. Total length is 5½ to 7½ inches (139 to 190 mm) and the tail is 1½ to 2⅛ inches (39 to 54 mm) in length. Adult weights range from 1¼ to 1⅞ ounces (34 to 54 gm). Males are slightly larger than females.

Distribution and Abundance. The meadow vole is primarily a northern species whose range extends down the Atlantic coast into South Carolina. It is abundant throughout Maryland and Virginia, but its distribution becomes less uniform farther south. In North Carolina it is locally common in the piedmont and mountains, uncommon to rare in the northern coastal plain, and absent from the southern coastal plain. It has been recorded only from the western piedmont and mountains and from Charleston County on the coast in South Carolina.

Habitat. The name meadow vole is most appropriate, as this animal prefers damp meadows; it also occurs in

coastal brackish and salt marshes, grassy upland fields, and orchards with a dense layer of herbaceous plants covering the ground.

Natural History. Meadow voles are active both during the day and night, spending most of their time in runways that they build beneath dense vegetation. They feed on the leaves and stems of a variety of grasses and forbs, often leaving small piles of cuttings scattered along their runways. They also eat fungi and insects and sometimes consume other animal matter. When their numbers are high these mice may cause damage to trees by eating the bark.

Reproduction probably occurs throughout the year, except for the coldest month or two in more northern parts of the region; breeding activity is reduced in the summer months. Four to 6 young usually make up a litter. The young, born naked with eyes and ears closed after a 21-day gestation period, grow rapidly and are weaned at about 14 days. Sexual maturity is attained at about 25 days, and females produce several litters each year, making this the most prolific rodent in the region.

These abundant animals are important prey for predators such as foxes, weasels, hawks, owls, and snakes. A high reproductive capability and rather broad habitat requirements assure their survival, but they may exhibit great fluctuations in numbers, being very abundant one year and difficult to find the next. When abundant, they can consume crops and damage orchards, but they are a staple food item for many predators and are thus important components of com-

plex food webs that characterize many communities.

Rock Vole
Microtus chrotorrhinus

Description. This mountain-dwelling vole is similar in size and appearance to the more common and widely distributed meadow vole. The upper body, however, is darker (blackish brown) and the snout is yellowish orange. Underparts are dark plumbeous gray. Adults range from 5⅝ to 7 inches (144 to 177 mm) in total length, and the tail is usually 1¾ to 2⅛ inches (46 to 55 mm) long. Weight of adults is usually 1 to 1¾ ounces (30 to 48 gm).

Distribution and Abundance. An isolated population of this rare rodent has been found in the Appalachian Mountains of North Carolina, eastern Tennessee, and West Virginia; this population is disjunct by about 250 miles (400 km) from the remainder of its range, which extends from northeastern Pennsylvania northward into eastern Canada. Records from western North Carolina are from areas above 3,800 feet (1,158 m). This rodent was probably more widespread in the southern Appalachians in the past, but now occurs only in isolated pockets of suitable habitat, such as the spruce-fir forests in the Great Smoky Mountains.

Habitat. This species is usually associated with rocky habitats either within high mountain forests or open fields, where it tunnels beneath a thick layer of leaf litter, or under mossy rocks and logs.

Rock vole (Microtus chrotorrhinus). *Photograph by Roger W. Barbour.*

early spring to late fall and 1 to 7 young are born after a gestation period of 19 to 21 days. Bobcats, rattlesnakes, copperheads, and northern short-tailed shrews are known to prey on rock voles, but surely there are other predators as well.

Natural History. The rock vole seldom is seen in this region, even by professional mammalogists, so few details are known about its biology. It eats many types of herbaceous plants, and occasionally consumes insects and fungi. Rock voles reproduce from

Woodland Vole

Microtus pinetorum

Description. Woodland voles are somewhat smaller than meadow voles and are adapted for a fossorial existence. They have a smooth dense coat of short chestnut-brown fur on the upperparts with the underparts a silvery gray. The eyes and ears are reduced in size. Although the general body shape is similar to that of other voles, woodland voles have a very

Woodland vole (Microtus pinetorum).

short tail and enlarged front claws adapted for tunneling. Total length of adults is usually 3⅞ to 5⅛ inches (97 to 131 mm), with a tail of only ¾ to 1⅛ inches (18 to 30 mm). Body weight is usually ⅝ to 1 ounce (17 to 30 gm). Woodland voles from the Carolinas and southern Virginia have smaller feet and brighter pelage than those from farther north.

Distribution and Abundance. This mouse is widely distributed in the eastern United States and occurs throughout this region. It is common in Maryland and Virginia, whereas in North Carolina it varies from being uncommon in some localities to common in others. It appears to be widely distributed in South Carolina but seldom abundant.

Habitat. These mice are adapted for burrowing and spend most of their lives in extensive systems of tunnels that lie just beneath the surface of the ground. They occupy a variety of woodland and old-field habitats, especially those with well-drained soil and

either a deep layer of leaf litter or dense vegetation on the ground. Woodland mice often find appropriate habitat in orchards and gardens.

Natural History. A variety of plant materials is consumed, primarily the shoots of grasses and forbs. In the autumn roots and seeds also are eaten, and in winter they feed primarily on bark and roots. In addition, they readily consume potatoes, fallen apples and pears, and the underground portions of nursery stock. They may completely girdle trees in the process of feeding on bark.

Reproduction may occur at any season in the warmer portions of the Carolinas, but usually ceases during winter in the colder portions of the region. The gestation period is 20 days, and the young are weaned by 21 days of age. One to 5 young constitute a litter, with 2 to 3 being most common. Maturity is reached at about 3 to 4 months, and females may produce 1 to 4 litters each year.

These mice lack the high reproductive potential of meadow voles, but probably have fewer natural enemies capable of preying on them in their underground burrow systems. They are taken by foxes, hawks, owls, and other predators when they emerge from their burrows.

Woodland voles may do considerable damage to fruit orchards and such tubers as potatoes. Gardeners who find root crops gnawed and burrows along garden rows should usually attribute the damage to these mice and not to the carnivorous moles which are sometimes blamed.

Muskrat

Ondatra zibethicus

Description. This semiaquatic mammal is easy to identify by the presence of a laterally compressed, sparsely haired tail. The general appearance is of a rabbit-sized, rather heavy-bodied rodent with dark brown to nearly black fur. The pelage consists of relatively long guard hairs overlying a soft, thick, waterproof underfur. The ears are very short and nearly hidden by the fur, and the eyes are small. Each front foot has 4 toes which are not webbed. The hind feet are larger and each has 5 toes which are partially webbed and have a lateral fringe of stiff hairs. Total length ranges from 16⅛ to 28 inches (41 to 71 cm) and the average weight is about 2¼ pounds (1 kg), with males being slightly larger than females. Muskrats from the piedmont and coastal plain of Maryland, Virginia, and North Carolina are slightly darker in color and larger in size than those from elsewhere in the four-state region.

Distribution and Abundance. Muskrats are found in all states within the region but may be locally absent from sizable areas. Muskrats are abundant in the marshes surrounding Chesapeake Bay and in northeastern North Carolina. They are less abundant inland, rare along the coast of southeastern North Carolina, and absent from coastal South Carolina.

Habitat. This rodent seems to prefer brackish marshes dominated by bullrushes or cattails, but is likely present in most well-vegetated fresh or brackish marshes. Their presence is marked by the occurrence of their houses, large mounds of vegetation

Muskrat (Ondatra zibethicus).

which conceal the den. Muskrats are also present in many farm ponds and along streams. In these habitats they usually do not build houses, but rather build their dens in tunnels in the pond or stream bank.

Natural History. Muskrats are generally nocturnal but may be occasionally seen during the day. They are herbivorous, eating a variety of semi-aquatic plants, primarily bullrushes and cattails. Stems, leaves, and underground portions of plants are eaten in summer, but in winter underground or underwater parts become more important. Muskrats also eat many kinds of field crops, and occasionally animal matter such as shellfish is consumed.

Breeding occurs throughout most of the year in this region, and 2 or more litters containing 3 to 7 young may be produced each year. The young are born blind, naked, and nearly helpless after a gestation period of about 30 days. They are cared for by the mother and remain in the den for the first 2 to 3 weeks. They are weaned during the fourth week, and are mature and ready to breed by the time they are a year old.

Mortality is high and the average lifespan is short; many animals eat muskrats, including mink, raccoons, alligators, largemouth bass, owls, and marsh hawks. In addition, they are trapped for their fur and to a lesser extent for food. During the winter of 1975–76 over 8 million muskrats were trapped in North America; the fur from those animals was valued at over 30 million dollars, making the muskrat the most important North American furbearer. In spite of heavy natural predation and trapping, the high productivity of this species has enabled it to prosper in most areas where adequate habitat has been maintained.

Southern Bog Lemming
Synaptomys cooperi

Description. This vole has a robust body, a broad head with small eyes, a blunt nose, and short ears that barely extend beyond the relatively long, shaggy fur. The total length ranges from 4⅝ to 5⅜ inches (119 to 136 mm), including an unusually short tail of ⅝ to 1⅛ inches (17 to 27 mm); adults average about ⅞ to 1¼ ounces (26 to 36 gm) in weight. The upper pelage is a mix of black, gray, and yellowish brown and is a somewhat grizzled brown in appearance; the underparts are grayish white, with no sharp line of separation from the darker coloration above. The tail and feet are grayish or brownish black. Another

Southern bog lemming (Synaptomys cooperi).

distinctive feature of this vole is the presence of a shallow longitudinal groove on the front of each upper incisor.

Distribution and Abundance. There are 2 races, or subspecies, of southern bog lemmings in the region. The most widely distributed of these occurs throughout Maryland, the mountains and upper piedmont of Virginia, and the mountains of North Carolina; it tends to be localized and uncommon throughout this area. The second subspecies is present only in the vicinity of the Dismal Swamp in extreme southeastern Virginia and northeastern North Carolina. This small, localized population, known only from the 1895–98 period, was considered to be extinct until the capture of additional specimens in 1980. Recent extensive study has revealed the widespread occurrence of south-

ern bog lemmings in the Dismal Swamp area, and this race is not considered to be threatened. The Dismal Swamp race, shown in the accompanying photograph, differs from the mountain race in being somewhat larger and slightly brighter in coloration.

Habitat. The southern bog lemming inhabits sphagnum bogs, moist meadows, canebrakes, borders of marshes, and various types of grasslands and weedy fields; however, it is also found in moist woodlands, thickets, and orchards. Its principal food is green, succulent grasses and sedges, so the presence of these plants is the chief requirement of any habitat. A habitat need not be extensive in area, as these voles can live in small colonies in habitats too small to support other similar species, such as the meadow vole.

Natural History. Surface runways are constructed through thick and matted grasses and other vegetation; runways of other small mammals also are used, especially those built by voles of the genus *Microtus*. Both southern bog lemmings and meadow voles leave fecal droppings and piles of grass cuttings along their runways; the presence of lemmings is revealed by bright green or yellowish droppings in contrast to the dark brown or blackish ones of meadow voles.

In addition to runways, there usually is an underground tunnel system. A nest of grasses and leaves, about 3½ to 6 inches (89 to 152 mm) in diameter, usually is constructed within a chamber of the tunnel. Occasionally, however, the nest is at the surface, concealed by thick grass, a brush pile, or some other cover. This animal tends to be most active in early morning and early evening.

Breeding occurs at least from early spring until late fall. The gestation period is about 23 days and litters of 2 to 5 offspring are usual, though litter size may vary from 1 to 7. Young are weaned after about 3 weeks. A female normally produces several litters per year.

Southern bog lemmings are preyed upon by hawks, owls, weasels, foxes, and snakes. Perhaps a greater threat to their survival are the changes that occur in the bogs, marshes, and other habitats that are related to human activities.

Black Rat
Rattus rattus
(Introduced)

Description. The black rat is similar in appearance to the Norway rat, but is smaller and more slender. In addition, the snout is more pointed, the naked ears are longer and larger, and the tail has smaller scales and is longer than the head and body length. The animal is grayish brown or black above, darker in the midline than on the sides; the belly is light gray, yellowish, or white. The tail is solid gray, not bicolored as in the Norway rat. The fur generally is soft, but mixed with long, bristly hairs. This rat has a total length of 13¾ to 17⅜ inches (35 to 44 cm), including a tail of 7½ to 9⅞ inches (19 to 25 cm). Its weight ranges from 4 to 12⅜ ounces (115 to 350 gm).

Distribution and Abundance. Introduced into North America by 1609, the black rat apparently was distributed widely in the eastern portion of the United States by the latter part of the eighteenth century. The larger and

Black rat (Rattus rattus).

more aggressive Norway rat reached the continent in about 1775 and gradually drove the black rat from much of its range. The present distribution of the black rat is restricted primarily to the south Atlantic and Gulf coasts, as well as along the Pacific Coast. In the Carolinas, Virginia, and Maryland, populations of black rats are most likely to be found in the vicinity of shipping ports, such as Baltimore, Norfolk, Wilmington, and Charleston, but also may persist at scattered inland localities.

Habitat. Black rats are basically arboreal animals; they live and nest in trees, shrubs, or vines, and seldom burrow in the ground. Where they inhabit human dwellings, outbuildings, or storage facilities, they most likely live in attics, walls, and ceilings. Where black rats and Norway rats live together, they are separated ecologically, with black rats occupying the upperstories of buildings while Norway rats prefer lower levels, basements, sewers, or burrows.

Natural History. These rats are active primarily during the late afternoon and night and feed on a variety of grains, fruits, and vegetables, as well as animal matter. Nests usually are near a source of food or a food cache, and made of a variety of available materials, such as grass, paper, or rags.

Breeding begins by the fourth month after birth and mating occurs throughout the year, though most commonly from January to June. A female may produce 5 to 6 litters each year, with each ranging in size from 4 to 11 offspring and averaging 7. The gestation period is about 21 days. Females mate soon after giving birth; therefore, they often are pregnant while nursing a litter. The young are weaned after about 3 weeks.

Norway Rat
Rattus norvegicus
(Introduced)

Description. This robust rodent has a total length of 12⅝ to 18⅞ inches (32 to 48 cm), including a tail of 5⅞ to 8⅝ inches (15 to 22 cm), and a weight of 10⅝ to 19 ounces (300 to 540 gm). Its snout is somewhat elongate but blunt, and the tail is sparsely haired, with pronounced scaly rings. In contrast to the black rat, the naked ears of the Norway rat are not sufficiently long to reach forward over the eyes, and its tail is slightly shorter than the combined length of head and body. The pelage is rather coarse and short, grayish brown above and pale gray or yellowish white below. The tail is bicolored, being darker above; however, the line of separation is not always distinct. There may be variation

in color among wild Norway rats, with some tending toward black. Domesticated strains, developed for laboratory use, are albino or spotted.

Distribution and Abundance. The Norway rat arrived on the east coast of the United States about 1775 and on the Pacific Coast about 1851. It subsequently has become distributed throughout much of North America, usually displacing the black rat. It occurs throughout the Carolinas, Virginia, and Maryland. Numbers vary according to habitat, being exceedingly abundant in poorly-kept or unclean residential, farm, or business areas.

Habitat. This rat most often lives in proximity to humans, wherever it can find food and shelter: the ground floors of buildings, tunnels or sewers beneath buildings, wharves, garbage dumps, and storage bins or elevators. In rural areas, especially during summer months, it may invade cultivated fields, pastures, and disposal facilities, and even inhabits salt marshes and barrier islands along the coast.

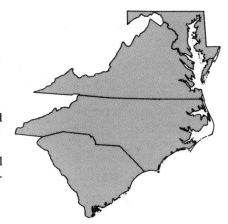

Norway rat (Rattus norvegicus).

Natural History. Norway rats are active primarily at night; during the day they occupy nests made of shredded rags, paper, or various types of debris, and located in the walls, floors, or foundations of buildings, or in underground tunnels. A tunnel system usually has several entrances, under boards, rocks, or a shallow soil cover. Chambers about 6 inches (15.2 cm) in diameter serve as nest sites or for food caches. These rats are colonial, with 10 to 12 individuals typically living together cooperatively, dominated by an older, large male.

Almost anything edible is included in the diet, including organic garbage of all kinds, grains and other plant materials, and flesh. They often kill other animals, including poultry, birds, rabbits, black rats, and sometimes their own young. They, in turn, are prey for snakes, carnivorous mammals, and many birds of prey.

Norway rats are prolific breeders year round; most young are produced in spring and fall. A female may give birth to 6 to 8 litters per year, with 7 to 11 young each. The gestation period is 21 to 23 days. Young are born naked, blind, and helpless; the eyes open in 14 to 17 days, and the young are weaned after about 3 weeks. Sexual maturity is reached at 3 to 5 months of age. Individuals may live for 2 to 3 years.

The most effective control measure against these introduced rats is to deny them suitable habitat and food sources. Other measures, such as trapping, shooting, and use of poison, are less effective and can have undesirable secondary consequences.

House Mouse

Mus musculus
(Introduced)

Description. This mouse is so familiar to most people that little description is required. It is small, the total length being 5½ to 7⅛ inches (140 to 180 mm), including a tail of 2¼ to 3⅝ inches (58 to 93 mm); it weighs from ⅝ to ¾ ounce (18 to 23 gm). It has an elongate snout; small, slightly protruding black eyes; large, naked ears; and a scaly tail that is scantily haired. Color varies from grayish brown to brown above, grading to gray, buff, or white on the belly, feet, and the underside of the tail.

Distribution and Abundance. The house mouse was introduced from Europe probably during the years of the American revolution. The species oc-

curs throughout the Carolinas, Virginia, and Maryland and is abundant in suitable habitat.

Habitat. Like the Norway and black rats, the house mouse is found commonly in close association with humans—in houses, restaurants, factories, warehouses, barns, or other man-made structures where food and space are available. It also lives abundantly in the wild, in weedy and overgrown abandoned fields, fencerows, grain fields, sand dunes, and other similar natural habitats. Nests are made of shredded paper, fabric, grasses, or other soft materials and are hidden in walls, under floors, trash piles, logs and stones, burrows, and in other equally protected places. The presence of house mice in a particular habitat usually becomes evident from their sign of shredded nesting

House mouse (Mus musculus).

material, damage caused by gnawing, and abundant spindle-shaped fecal droppings.

Natural History. House mice are active year round, mostly at night, but activity is reduced in winter when they tend to remain in nests, presumably to conserve heat. These mice are relatively sedentary; however, individuals sometimes travel over considerable distances between winter quarters in buildings and outdoor nest sites used in spring and summer. They are highly social and form small colonies, usually consisting of a male and several females with their young. Being aggressive in behavior, colony members drive other mice from the nest area. House mice tend to compete successfully with native mice when they invade favorable habitats in the wild.

House mice consume anything edible. They are highly destructive to stored grains and other foods eaten by humans and domestic animals. In the wild, they eat weed seeds, insects, and a variety of other plant and animal matter. They are preyed upon by domestic cats, rats, and various carnivorous mammals, snakes, and birds of prey.

Like most rodents, house mice are prolific. Breeding extends mostly from early spring to late fall, and may occur year round, producing as many as 13 litters in a year. Litter size varies from 3 to 12 young, with an average of 6. The blind, naked, and helpless young are born after a gestation period of 19 to 21 days. Fur is present after 10 days and the eyes open by 14 days; they are weaned after 3 weeks and become sexually mature at 6 to 8 weeks.

Due to their high reproductive rate and adaptability, these destructive little mammals may require control. Snap traps baited with rolled oats, peanut butter, bacon, or other foods usually are effective. Poisons should be avoided.

Meadow Jumping Mouse
Zapus hudsonius

Description. The total length of meadow jumping mice is 7⅝ to 8½ inches (194 to 215 mm); over half of that length is an extremely elongate tail of 4⅜ to 5¼ inches (110 to 132 mm). These mice also have well-developed and incredibly long hind legs and oversized hind feet, in contrast to relatively short front legs; the hind foot measures 1 to 1⅛ inches (26 to 30 mm) in length.

The short, rather coarse pelage of this attractive mouse is distinctively colored. Middorsally, a dark yellowish

Meadow jumping mouse (Zapus hudsonius). *Photograph by John R. MacGregor.*

brown band, interspersed with numerous black-tipped hairs, extends from the nose to the base of the tail; the sides are yellowish orange, also with some of the hairs tipped with black; the underparts and feet are white, sometimes suffused with yellowish orange. The tail is bicolored, brownish above and white below, sparsely haired, and with a slight tuft of dark hairs at the end.

The species can be separated readily from other mice in the region, except the woodland jumping mouse, on the basis of color and length of the tail and hind legs and feet. It differs from the woodland jumping mouse in having less bright colors, and the tuft of hair at the end of the tail is dark rather than white.

Distribution and Abundance. This species occurs statewide in Maryland and Virginia, in the piedmont and mountains in North Carolina, and in the westernmost counties in South Carolina. In the mountains, at least in North Carolina, it is found up to elevations of 3,000 feet (914 m). The meadow jumping mouse is uncommon

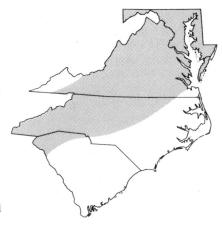

or rare over most of its range in the region and has been reported as abundant only locally, as on Assateague Island in Maryland and in certain areas proximate to the Blue Ridge Mountains in Virginia.

Habitat. Meadow jumping mice usually are found in moist weedy or grassy fields and in thick vegetation near marshes, streams, or ponds. Though they most often are associated with open areas, they sometimes inhabit woodlands with a growth of herbaceous plants, especially where woodland jumping mice are not also present.

Natural History. Jumping mice are the only mice in this region which hibernate; they enter hibernation in late November or December and emerge in late March or early April. They do not store food, but accumulate fat prior to winter; an animal that weighs half an ounce in May or June may weigh almost an ounce in September or October. They may hibernate singly or in small, closely huddled groups in nests of grass, leaves, or other insulating vegetation in underground burrows that may be as deep as 3 feet (0.9 m). A dormant animal usually rolls up with the head between the hind legs and the tail curled around the body. Should the weather moderate, individuals may emerge and become active for a brief time.

This lively mouse is active primarily from early dusk until dawn. It wanders freely, foraging among the weeds and grasses for insects, seeds, nuts, berries, fungi, and fruits. Well-defined trails or runways seldom are visible. It usually moves rather slowly through the grasses or takes short hops rather than progressing by jumping. When startled or in retreat, a mouse may take several jumps of a foot or more and then freeze in place; its protective coloration is an aid in avoiding detection in grassy habitats. Maximum leaping distance is about 3 feet (0.9 m).

A summer nest of grass and leaves is constructed in a protected location such as in a hollow tree or log, under logs or clumps of vegetation on or above the ground, or in burrows. Breeding commences soon after the animals emerge from hibernation in the spring and continues until fall. Usually 2 litters are produced each year, with 3 to 8 young born per litter; the gestation period is about 18 days.

These mice are essentially solitary in nature and docile in behavior. They seldom come into contact with humans and unfortunately are seen rarely. They do have numerous enemies, however, including snakes, several carnivorous mammals, and raptorial birds.

Woodland Jumping Mouse
Napaeozapus insignis

Description. This extremely energetic mouse is 8⅜ to 9 inches (213 to 230 mm) in total length and has elongate hind limbs, the hind foot measuring 1⅛ to 1¼ inches (29 to 31 mm) in length. The long, scaly tail is 5⅛ to 5¾ inches (130 to 146 mm) in length, and is grayish brown above and white below except for the tip which is totally white. The dorsal pelage of the

Woodland jumping mouse (Napaeozapus insignis).

woodland jumping mouse contains a mixture of reddish orange and black hairs that form a distinct broad band down the back. The sides are yellowish orange and the belly is white.

Distribution and Abundance. The woodland jumping mouse is locally abundant in the mountains of Maryland and Virginia, but uncommon in North and South Carolina, where it is restricted to suitable habitats above 2,800 feet (853 m) in elevation.

Habitat. Cool, moist spruce-fir and hemlock-hardwood forests of the high mountains are the favored haunts of this species, especially along the banks of streams and lakes in areas with low ground cover. It also inhabits bogs, swamps, and damp rocky seeps, but seldom occurs in open fields and meadows as does the meadow jumping mouse.

Natural History. This mouse is primarily nocturnal, but may also be active at dusk and dawn if the weather is rainy or overcast. Although able to jump several feet in a leap, it usually takes a few moderate hops and then stops abruptly under cover and re-

mains motionless unless pursued. Woodland jumping mice do not construct runways, but utilize and modify passages excavated by other rodents, shrews, and moles; they also climb well in bushes and other low vegetation.

The woodland jumping mouse accumulates a large reserve of fat in the autumn and then hibernates from November to April in this four-state region. Nests, the entrances to which are closed during the day, are built of grasses and dried leaves in an underground chamber or in piles of brush; the mice also utilize fallen trees and rotting logs as shelter.

Two litters are born in the summer, the first in June or July followed by another in August or September. Two to 7 (usually 4 or 5) young are born after a gestation period of 3 to 4 weeks. The young are born pink and naked, and the eyes and ears are closed, but by 4 weeks of age the body is well furred and the eyes and ears are open. The young are weaned at about 5 weeks of age. Males and females become sexually mature after the first year.

Woodland jumping mice feed on fungi, seeds and other plant material, insects, grubs, and worms. Animals known to prey on these mice include striped skunks, mink, bobcats, feral cats, screech owls, timber rattlesnakes, and copperheads, but other predators surely exist. The woodland jumping mouse does poorly in captivity unless adequate material is provided in which it can hide.

Porcupine
Erethizon dorsatum
(Extirpated)

Description. The porcupine is a robust, somewhat dumpy animal with a small head and short legs and tail. Its feet have broad, naked soles and long, curved claws. Typical adults may weigh 8⅞ to 26½ pounds (4 to 12 kg), though large males may exceed 37½ pounds (17 kg). Their total length averages about 33½ inches (85 cm), including a tail of about 9 inches (23 cm).

The pelage of porcupines includes hairs modified into stiff, sharp-pointed, hollow quills that may measure 3 inches (76 mm) or more in length. These unusual structures are present over the upper parts of the body, but not on the underside. The quills are distributed among longer, coarse guard hairs and dense, woolly underfur, giving the animals a shaggy, ungainly appearance. The pelage is dark brown to blackish, and the quills are yellowish white at their base.

Distribution. The present distribution of the species in North America extends from the edge of the Canadian tundra southward along the Rocky Mountains to Mexico. Early records indicate that the porcupine was present in the eighteenth and nineteenth centuries in parts of Maryland and Virginia, and reports of the animal in western Maryland and West Virginia have persisted until recently; it is likely that such reports are of wandering animals. It seems certain that the porcupine was extirpated from this portion of its original range and that

Porcupine (Erethizon dorsatum).

at present the southern limit of its distribution in the east is south-central Pennsylvania.

Habitat. Porcupines are primarily creatures of mixed conifer-hardwood woodlands upon which they depend for their principal food supply.

Natural History. "Quill pigs," as porcupines are often called, are primarily vegetarians. Their winter diet is almost exclusively foliage and soft tissues beneath the outer bark of a wide variety of woody shrubs and trees. They feed upon almost any conifer in their range, as well as such hardwood species as maple, oak, beech, and birch. Trees often are deformed, or killed by girdling, as a result of being climbed and gnawed by porcupines. In spring and summer, buds and twigs are eaten, as are herbaceous vegetation and mast.

Porcupines are inoffensive, but they defend themselves effectively with their arsenal of quills, which are loosely attached to the skin and bear tiny overlapping barbs at their tips. If threatened or attacked, an animal turns its back to its foe, raises the quills, and lashes its tail from side to side. Quills are not thrown by porcupines as is popularly believed, but they do detach easily to become embedded in the flesh of the attacker. The quills are difficult to remove and usually work progressively deeper into the victim and can cause death. Some predators, such as bobcats, coyotes, mountain lions, and fishers, succeed in killing porcupines, apparently by attacking their underside, which is not protected by quills.

Porcupines are nocturnal and active year round. Much of their time is spent foraging in trees or on the ground. The den usually is in a hollow tree, under a log or rocky ledge, or in a cave; nest material is not collected. These animals are generally solitary, except for mother and young, but in winter a den may be shared by a group.

Mating occurs in fall and is initiated by elaborate courtship behavior. Copulation is accomplished as the male sits upright, not leaning over the female, and the female turns her amply-quilled tail aside. A single precocious offspring is born between April and June after a gestation period of 7 months. The infant at birth is about 10 inches (254 mm) long and weighs about 1⅛ pounds (0.5 kg); its eyes are open, its teeth are well formed, and hair is present, including soft quills which harden within an hour. It is able to eat solid food at once, though it nurses for several months. It can

climb and assume a defensive posture soon after it is born. Young porcupines grow rapidly and weigh from 3⅛ to 4 pounds (1.4 to 1.8 kg) in 4 or 5 months.

Nutria

Myocastor coypus
(Introduced)

Description. This large rodent resembles a very large muskrat. The tail of the nutria, however, is round and nearly naked, and the fur is coarse and ragged in appearance. The color of the hair varies from dark brown to yellowish brown with the belly typically being a pale gray; the muzzle is usually grayish. The fur consists of long outer guard hairs and short soft underfur. The hind feet are webbed.

Nutria (Myocastor coypus).

Nutria are large rodents; adults from Louisiana may be as much as 3¼ feet (1.0 m) in total length and usually weigh 7⅞ to 20 pounds (3.6 to 9.1 kg).

Distribution and Abundance. The nutria is native to South America but has been introduced into several North American localities including Virginia and North Carolina. It is present along the coast in Maryland, Virginia, and North Carolina south at least to Morehead City. It also occurs in the Chesapeake Bay and Currituck and Pamlico sounds. There are inland records in North Carolina from Rockingham County and in Maryland from Garrett County.

This exotic species does well in the coastal marshes of this region and numbers are high in many places. It appears to be moving southward in the coastal marshes of North Carolina and may also be extending its range inland from the brackish marshes of Pamlico Sound into nearby freshwater systems.

Habitat. Nutria occur in smooth cordgrass salt marshes, brackish waters of coastal estuaries, as well as in some freshwater marshlands.

Natural History. This rodent is primarily a nocturnal herbivore that feeds on a variety of marsh plants, such as three-square rushes, cordgrasses, and cattails. The stems and underground portions of these plants are consumed. They often build large feeding platforms of vegetation which may appear much like flattened muskrat houses. Nearby plants are cut and carried to the platforms to be eaten. Nutria sometimes move from marshes into nearby agricultural fields to feed on crops such as grains.

Nutria apparently reproduce throughout the year, and females begin breeding during their first year. The gestation period is from 127 to 132 days; 4 to 6 young usually comprise a litter, and 2 litters are often produced in a year. Young nutria are born fully furred and precocial; they are usually weaned during the second week of life. Individuals are very sedentary and home ranges are small. Nutria build complex burrow systems with several nest chambers that are used for shelter by adults and young.

Along the Gulf Coast, alligators regularly take nutria, and nutria remains have been found in the nests of bald eagles around the Chesapeake Bay. Although adults are probably safe from most smaller predators, the young likely are taken by most of the same predators that feed on muskrats. Nutria also suffer from parasites and disease, and sometimes are killed by severe winter weather.

Nutria may be considered either a nuisance or a resource. They sometimes damage crops growing near marsh habitats, and their burrow systems may penetrate dikes and drain ponds or cause the collapse of roads. At high densities, they may consume so much marsh vegetation that they convert marsh habitat to open water habitat. On the other hand, nutria are harvested by trappers, and in some southern states they are considered a valuable fur resource.

Meat-eating Mammals
Order Carnivora

Even though some mammals in other orders are flesh eaters and a few members of this order utilize food items other than flesh in their diets, the Order Carnivora includes some of the most efficient, specialized, and highly adapted predators among mammals. Their structural, functional, and behavioral characteristics are focused upon seeking out, killing, and consuming prey organisms. They are vital components of natural ecosystems, helping to maintain the health and stability of populations of numerous prey species.

This large and diverse order is cosmopolitan in distribution. A total of 19 species has been reliably reported from the Carolinas, Virginia, and Maryland. Many, such as the raccoon and foxes, are distributed widely in the region; others, such as some of the weasels, are more limited geographically. The status of the coyote and mountain lion remain uncertain. The carnivores of the region are in 6 families.

Family Canidae includes 5 species of doglike carnivores: the red and gray foxes, red and gray wolves, and the coyote, which have elongated muzzles, long legs, and prominent bushy tails.

Family Ursidae contains the black bear, the largest carnivore in the region. Bears are characterized by plantigrade feet and a short tail and are more omnivorous than most other carnivores.

Family Procyonidae includes the raccoon, a medium-sized, somewhat omnivorous carnivore that is readily recognized by its ringed tail and black bandit-style mask across the eyes.

Family Mustelidae is highly diverse, with 7 species well documented from the region, including weasels, skunks, and the mink, fisher, and river otter. In addition, the marten (*Martes americana*) and ermine (*Mustela erminea*) have been reported from Washington, D.C., and Maryland, respectively; however, these records are outside the normal ranges of these species, and probably do not represent established natural populations. Mustelids generally have an elongate, slender body, and paired, often potent, musk glands in the anal region.

Family Phocidae includes the hair seals, a group adapted to life in the sea and along its shores; 3 species have been reported along the coastline of the Carolinas, Virginia, and Maryland. Their feet and limbs are modified as flippers, and their body is streamlined with the knees and elbows enclosed within the body contour.

Family Felidae includes the cats, characterized by well-developed canine teeth and retractile claws. Only the bobcat and mountain lion are native to this region.

Most carnivores other than seals are terrestrial, and except for bears and raccoons have digitigrade feet, walking on their toes, with the heel raised off the ground. Digits are equipped with strong claws. Jaws are equipped with large, conical, usually recurved canine teeth. The cheek teeth of most species are designed for cutting and shearing flesh; the last upper premolar and the first lower molar teeth are well developed and bladelike, and do most of the shearing. The jaw muscles are powerful, and jaw movement is in the vertical plane, insuring an effective bite. These adaptations for a carnivorous diet contrast dramatically with those of herbivores which have flattened, broad cheek teeth and jaws that move both vertically and laterally to crush and grind. Bears and raccoons have a varied diet that includes vegetation; their teeth tend to be intermediate between those of other carnivores and the herbivores.

Carnivores have a keen sense of smell and sight, enhancing their ability to locate prey; most see well in the dark or in dim light and, hence, are active at night. Many of the larger carnivorous species, such as wolves and cats, feed relatively infrequently, depending upon availability of prey and hunting success. They often experience famine between kills, but are able to ingest large quantities of food when a kill is made, up to 20 pounds (9 kg) by a wolf. A well-developed foregut and a large simple stomach make this possible. In contrast to that of herbivores, the hindgut is short, for food is digested quickly.

Coyote
Canis latrans

Description. The coyote is a doglike mammal with a body, tail, and pointed snout much like that of a fox. The coyote holds its tail down when running in contrast to foxes, which tend to hold their tails horizontal to the ground. Body color ranges from gray to reddish brown with the extremities being rusty brown. Total length of coyotes from Indiana ranges from 3 feet 4 inches to almost 4 feet (1 to 1.2 m). Weight is usually between 20 and 44 pounds (9 to 20 kg) with females usually 2¼ to 4⅜ pounds (1 to 2 kg) lighter than males. Individuals are often extremely difficult to distinguish from some dogs, and hybrids do occur.

Distribution and Abundance. Coyotes are generally considered to be animals of the western plains. However, in recent years, as large areas of eastern forests have been cleared, this opportunistic animal has moved east, sometimes with the help of man. They have been released in several eastern states and have been reported from all states in this region. Reports are relatively uncommon but are becoming more frequent and widespread throughout the region.

Habitat. This member of the dog family is usually associated with semi-open country. As eastern forests have been cleared and converted to farm land, the resulting patchwork of fields and brushy edges apparently has created suitable habitat. The open grasslands associated with interstate highways have probably provided corridors

Coyote (Canis latrans).

of suitable habitat extending from western grasslands into the four-state region.

Natural History. Coyotes are active predators and scavengers, taking a variety of prey, including rabbits, rodents, birds, and occasionally larger mammals such as deer. They also may consume significant amounts of carrion, especially during the winter, and the habit of feeding on dead carcasses of large animals is responsible for much of their notoriety as livestock killers. They tend to be most active during early morning and late afternoon but may also be active at night or during midday.

In the west, coyotes occur as lone individuals or family groups. Mating usually takes place in late winter or early spring. An average of 6 blind, helpless pups are born after a 63-day gestation period. Birth usually occurs

in a den which may be in a burrow, rockpile, or hollow log. Young coyotes are weaned at 5 to 7 weeks of age and are then fed by both parents and sometimes by helpers, which may be the mother's young from the previous year. Pups usually emerge from the den at about 3 weeks of age and disperse by winter.

Coyotes are affected by most of the same diseases as foxes. The major cause of mortality, however, is man. They are hunted and trapped; in addition, there are active control programs in many parts of their range due to their reputation as predators of livestock. In spite of efforts to reduce their numbers the coyote continues as an efficient scavenger and predator in many parts of the west and is becoming more widespread and abundant in the east.

Red Wolf

Canis rufus
(Extirpated)

Description. The red wolf is intermediate in size between the coyote and gray wolf. Specimens from Texas measure from about 4½ to 5½ feet (1.4 to 1.7 m) in total length. Males usually weigh 44 to 90 pounds (20 to 41 kg), whereas females weigh 35 to 64 pounds (16 to 29 kg). It has a more slender build than the gray wolf and also has longer legs. Some individuals are reddish in color; others, however, may be gray or even black.

Distribution and Abundance. This southern wolf once apparently extended northward along the east coast as far as southern Pennsylvania. The limits of its distribution and abundance in the past are poorly documented because it has been extirpated from the region at least since colonial times. In fact, due to continued persecution by humans and hybridization with an expanding coyote population, it is likely that genetically pure red wolves no longer exist in the wild. A few have been taken into captivity and the species is being maintained in breeding facilities for possible future reintroduction.

Habitat. The red wolf occupied habitats with large amounts of cover in both upland and swamp forests as well as coastal marshes and prairies.

Natural History. Little is known about the biology of this species. It was apparently most active at night and fed primarily on rabbits and rodents; it was, however, capable of taking larger prey.

Mating evidently takes place in early spring, and the females give

Red wolf (Canis rufus). *Photograph courtesy Texas Parks and Wildlife Department.*

birth to their pups in late spring after a gestation period of 60 to 63 days.

The red wolf declined sharply in numbers apparently due to a combination of factors including deliberate killing by man, habitat alterations, diseases, parasites, and competition from and hybridization with coyotes. If this federally Endangered Species exists at all in the wild, it is restricted to a few individuals in coastal Louisiana and Texas. The future existence of wild populations probably depends on the reintroduction of individuals raised in captivity into such isolated locations as southern coastal islands. Two such introductions occurred on Bulls Island, South Carolina. Mated pairs were released in 1976 and 1978, but these animals were subsequently recaptured and returned to captivity.

Gray Wolf
Canis lupus
(Extirpated)

Description. This is the largest of the wild dogs of North America. Gray wolves resemble large German shepherd dogs but have longer legs and bigger feet. The usual coloration is gray, but nearly white or black individuals sometimes occur. Gray wolves usually range from 4 feet 3 inches to almost 6 feet (1.3 to 1.8 m) in total length and stand 27½ to 31½ inches (70 to 80 cm) at the shoulder. Males weigh 44 to 176 pounds (20 to 80 kg), and females weigh 40 to 121 pounds (18 to 55 kg).

Distribution and Abundance. These wolves ranged throughout the eastern United States during early colonial

Gray wolf (Canis lupus).

times. They have, however, been totally eliminated from the east, the closest population being in upper Michigan. The last gray wolf was killed in North Carolina in Haywood County in 1887 and in Virginia in 1910 in Tazewell County.

Habitat. In regions where gray wolves still occur, they occupy a wide variety of habitats, from forests to plains. They appear to be limited more by the availability of suitable prey than by the physical characteristics of the habitat itself.

Natural History. Gray wolves hunt primarily in groups called packs for large prey such as deer. They also consume rodents and hares and may take livestock. Wolf predation is quite selective and often results in the culling of old, sick, and wounded individuals, as those are easiest to catch.

Breeding begins at 2 to 3 years of age, and some males and females remain together for several years. Courtship begins in late winter and the young are born the following spring in an excavated den after a 63-day gestation period. Litters average 6 young which are blind and helpless at birth. Pups are weaned at about 5 weeks, and by autumn are large enough to join the pack as it hunts for food. Gray wolves are highly social animals and there are strong bonds among pack members, including a very rigid social order in which a dominant male and female lead the pack.

The gray wolf is sometimes feared and hated because it may prey on livestock and is large enough to make some people feel threatened. Thus, it has been hunted, trapped, and poisoned and has been extirpated from most of its original range in the conterminous United States; it is classified as federally Endangered or Threatened in those states which retain remnant populations. Few easterners lament the loss of this magnificent predator; however, recent studies have helped to remove some of our misconceptions and prejudices regarding the wolf, and perhaps public attitudes towards large predators will change.

Red Fox
Vulpes vulpes

Description. The red fox looks much like a bushy-tailed, sharp-nosed, medium-sized dog. Its head and upper body are usually rich reddish yellow, but black or silver color phases also occur. The elongate bushy tail is tipped with white, and the feet and ear edgings are black. Red foxes are about 39 to 41 inches (100 to 103 cm) in total length and weigh 9 to 12 pounds (4.1 to 5.4 kg). Males average larger than females.

Distribution and Abundance. The red fox has expanded its range southeastward since colonial times and now occurs throughout most of the region; it is doubtful that this fox occurred here prior to that time. This species also has been widely stocked in the eastern United States with animals from other parts of North America and Europe.

Red fox (Vulpes vulpes).

The red fox is common in the mountains and piedmont, but rare to absent in the coastal plain. No specimens are known from the Dismal Swamp or Outer Banks.

Habitat. These foxes usually are associated with open habitats and are seldom found in dense woodlands. They prefer areas with interspersed croplands, woodlots, and old fields. They frequent the edges between different types of cover, and often place their dens in open fields.

Natural History. The red fox is perhaps best known as an efficient mouser, because mice make up a major portion of its diet, especially in winter. Cottontails are also an important food source, and in summer insects and plants are eaten. Red foxes are opportunistic predators, and many

other foods including birds and their eggs and other small mammals are utilized when available.

Red foxes are active throughout the year. They pair for life, usually breeding in late winter, with some individu-

als mating during their first year. After a gestation period of 52 days, 4 or 5 young are born usually in early spring in a den which may have been abandoned by a woodchuck or dug by the foxes. Both parents share the duties of caring for the pups. The male brings most of the food for both the young and the female until the pups can be left alone for short periods. At this time the female resumes hunting. The pups remain in the den for about a month before they begin to venture out into the open. By 3 months of age they begin to explore the area around the den, and by late summer or early fall they start to leave the den site and become independent.

The primary causes of death in foxes relate to man. Foxes are hunted, trapped, and killed by automobiles and farm machinery. They are also subject to a variety of diseases including mange, distemper, and rabies. Attitudes toward red foxes are highly variable in different parts of the range. There have been bounties on them at times, and some states consider them game or fur-bearing animals. Fox hunting has been a popular sport in this region since colonial times. They are usually hunted with hounds by hunters either on horseback or on foot. Often the chase is the primary object of the hunt and the taking of the fox is not necessary for the hunt to be successful.

Although foxes may still be viewed as harmful predators, people are beginning to recognize that they are valuable components of many natural ecosystems, and are very effective in the control of rodents. In spite of being hunted, trapped, and sometimes poisoned, foxes are likely to remain a viable part of our wildlife heritage for many years. They live up to their reputation as sly, adaptable mammals coping well with modern land use practices.

Gray Fox
Urocyon cinereoargenteus

Description. The dominant color of this fox is gray, as its name implies; however, the presence of reddish patches of fur on the sides of the neck, flanks, legs, and undersurface of the tail often causes it to be misidentified as a red fox by those not familiar with both species. The gray fox always has a black-tipped tail, whereas that of the red fox is white. Gray foxes are usually somewhat smaller than reds, being 33½ to almost 40 inches (85 to 100 cm) in total length and weighing about 7 to 10½ pounds (3.2 to 4.8 kg).

Distribution and Abundance. The gray fox occurs throughout much of this region. It is not found on the higher mountain slopes and is uncommon along the Eastern Shore, but it is abundant in other inland localities and on most coastal islands.

Habitat. These foxes live in habitats similar to those occupied by red foxes but appear to have a stronger preference for woodlands, especially those in the early successional stages of forest development. They readily live

Gray fox (Urocyon cinereoargenteus).

near humans and often occur in the suburbs of large cities.

Natural History. The gray fox, like its red cousin, depends heavily on cottontails and rodents for its food, but includes insects and such fruits as

grapes, apples, and berries in its diet, especially in summer.

The breeding season extends from January to April but begins somewhat earlier in more southern areas. After a gestation period of about 53 days, 4 (range 2 to 7) young are born. Their growth proceeds in a manner similar to that of the red fox, and they are usually independent by autumn. Both parents care for the young.

Gray foxes, like red foxes, are trapped, hunted, and otherwise subjected to mortality as a result of association with humans. Red and gray foxes also share many of the same diseases; however, gray foxes appear to present a greater danger in the spread of rabies than do red foxes. Rabid foxes may be unusually aggressive and attempt to bite humans and other animals, or they may appear sick and weak and make no attempt to bite.

Sick foxes or foxes that show no fear of humans should be avoided. In spite of this potential problem, gray foxes remain valuable as important agents in rodent control.

Black Bear

Ursus americanus

Description. The black bear is the only bear native to eastern North America. It is generally bluish black but occasionally may be brownish in color; there may also be a small patch of white hair on the chest. The ears and tail are very short and the eyes are small. Black bears walk on all 4 feet, with the entire surface of the foot placed on the ground. They vary considerably in size at maturity. Males

normally range from about 132 to 309 pounds (60 to 140 kg) but individuals have been reported to weigh in excess of 600 pounds (272 kg); females are smaller, averaging 88 to 154 pounds (40 to 70 kg). In this region, bears of the coastal plain tend to be slightly smaller than those of mountain habitats, and males average larger than females at any specific locality.

Distribution and Abundance. The black bear once occurred throughout eastern North America but now finds its range being restricted and its numbers decreasing. In the Carolinas and Virginia it occurs in both the coastal swamps and mountains, whereas in Maryland it occurs only in the mountains. The black bear wanders great distances, and reports of bears lost in cities are not uncommon. While bears

Black bear (Ursus americanus). *Photograph by Walter C. Biggs, Jr.*

or crude nests built in dense ever-green cover. Here they become inac-tive until the onset of warmer weather. Unlike true hibernators, however, denning bears may become fully alert if disturbed.

Females first breed at 3 to 5 years of age. Mating occurs during the summer, and the gestation period is 7 to 8 months. One or 2 blind and hair-less cubs are born usually during the period of winter dormancy. They re-main with the mother through the first year, dispersing during the sec-ond spring. Females, therefore, nor-mally breed only every second year.

Black bears generally live to about 10 years of age, although a few may survive twice that period. They have no major predators other than man. They are widely hunted, and most state conservation agencies operate management programs designed to maintain healthy populations. Hunting and habitat destruction appear to be the primary factors in determining population levels.

Although black bears are not gener-ally aggressive toward people, "park bears" may lose their natural caution and become dangerous if provoked. No bear should be approached, espe-cially those accustomed to the sight and smell of humans and to feeding on garbage. Use caution when around these large, strong, and sometimes unpredictable mammals.

are still common where protected, as in some state and national parks, they are relatively uncommon elsewhere. They face especially difficult times in coastal swamplands, as available habi-tat is being reduced by land clearing operations.

Habitat. In the mountains black bears prefer oak-hickory and mixed hard-wood forests, but in the coastal plain they are most often found in swamps, pocosins, and flatwoods. Although black bears are very adaptable, they appear to need large areas of refuge, generally with thick forest cover inac-cessible to humans.

Natural History. Black bears are omni-vores, eating many kinds of animal and plant foods. In spring they con-sume a variety of grasses and forbs, in summer and fall the diet emphasizes berries and fruit, and in winter acorns and other mast. The animal portion of the diet consists primarily of insects and grubs supplemented with carrion.

Most bears den up during the cold-est part of the winter in hollow trees

Raccoon
Procyon lotor

Description. The brownish black bandit mask, gray to brownish grizzled pelage, and conspicuous ringed tail make the raccoon one of the most easily identified mammals of this region. The fur is relatively long and tends to give raccoons a "roly-poly" appearance. Body color is usually gray in inland populations but may be brownish in those animals inhabiting coastal marshes. The forepaws are well adapted for manipulating objects and regularly are used in searching for and handling food items. Typical adult raccoons range from about 8 to 20 pounds (3.6 to 9.0 kg) in weight; males are usually about 10 to 15 percent heavier than females. Adults range from about 28 to 33 inches (72 to 84 cm) in total length.

Distribution and Abundance. The raccoon occurs throughout the Carolinas, Virginia, and Maryland. It is particularly abundant in the coastal plain but less plentiful in the piedmont and mountains.

Habitat. Raccoons are usually associated with wetland habitats such as marshes, swamps, and streams but also occur in moist upland habitats and even in suburban neighborhoods. Along the coast raccoons are abundant in the regularly flooded tidal marshes and are equally at home in forested swamps. Inland they tend to be associated with streamside forests but often wander into upland forests or agricultural lands.

Natural History. This animal is well known for its habit of washing food items before eating; however, biologists generally interpret this behavior

Raccoon (Procyon lotor).

as relating to feeling the food rather than cleansing it. The forepaws are well endowed with tactile sensors, and raccoons have considerable ability to discriminate among objects using their sense of touch. Their ability to use highly mobile and sensitive fingers to open latches on cages and remove garbage can lids is well known.

Raccoons are omnivores. They have a preference for crayfish and crabs, but also eat a variety of fruits, berries, and seeds. On coastal beaches raccoons become serious predators on the eggs of sea turtles, and they will eat the eggs of other animals when available. These hardy mammals are very adaptable and eat a wide variety of both plant and animal matter when preferred foods are absent.

Raccoons may become dormant in winter in those areas with a permanent winter cover of snow. In most of this region they are active throughout the year but may remain in their dens during periods of especially cold weather. Raccoons usually breed first during their first or second year and

mating usually occurs in late winter. Kits are born in a den in a hollow tree or burrow during late spring or early summer after a gestation period of 63 to 65 days. Litters usually contain 2 to 5 kits, which are cared for by the mother and weaned by 16 weeks of age. They remain with the mother, however, until they are about 9 months old. Although raccoons have been known to live for as long as 17 years in captivity, the life span of wild individuals is generally much less; some studies have shown average life spans of 2 to 3 years.

Many factors, including food shortages, parasites, and diseases, affect raccoon mortality. Individuals appearing sick and weak or aggressive and unpredictable should be avoided, for they may be rabid. Man has a major impact on raccoon populations, for they traditionally are hunted and trapped throughout the Southeast when long fur is in vogue.

Fisher

Martes pennanti

(Extirpated and reintroduced)

Description. The fisher is a fox-sized mustelid that is active mostly on the ground but sometimes in trees. It has a long body, relatively short legs, and a long bushy tail. The fur is dark brown above and nearly black on the belly. Streaks of gray or pale brown on the face and shoulders produce a grizzled appearance, and occasional white patches of fur may occur in the neck and throat region. Fishers are darker

Fisher (Martes pennanti). *Photograph by Roger A. Powell.*

in winter than in summer. Males are about 3 to 4 feet (90 to 120 cm) in total length and weigh 7¾ to about 12 pounds (3.5 to 5.5 kg). Females are much smaller than males, occasionally weighing half as much.

Distribution and Abundance. Fishers are generally associated with the mixed evergreen and deciduous forests of northern North America. They once occurred along the southern Appalachians at least to Virginia and early accounts seem to suggest their presence in the Great Smoky Mountains.

The fisher was reintroduced into West Virginia in the late 1960s and has subsequently been observed in the mountains of both Virginia and Maryland. Its present status in this region remains uncertain, but the species is considered Endangered in Virginia.

Habitat. Large stands of undisturbed northern hardwood-conifer forests are the preferred habitat of the fisher, but forested bogs also are inhabited.

Natural History. These opportunistic predators hunt both at night and dur-

ing the day, and take a variety of prey species such as rabbits, rodents, birds and their eggs, and, in the northern part of their range, porcupines. Although they hunt primarily on the ground, they are good climbers. They do not hibernate but may become inactive during periods of extreme cold.

Fishers are usually solitary during winter, but pair as the spring breeding season approaches. Mating occurs shortly after the young are born, and the gestation period lasts for nearly a year because the fertilized eggs remain quiescent for a prolonged period before actively beginning development; the period of active pregnancy is about 30 days. One to 4 young are born in March or April in a den, hollow tree, or other protected site. They are helpless at birth but grow rapidly; the eyes open after 8 weeks, and the young are weaned by 16 weeks even though they remain with the mother through summer.

Adult fishers apparently have no natural predators although the young may occasionally be taken by other carnivores or raptorial birds. They are trapped farther north where they occur in sufficient numbers. The pelts are highly valued in the fur trade, and trapping is closely monitored to prevent overharvest.

Least Weasel
Mustela nivalis

Description. The least weasel is the smallest carnivore in North America. Adult males reach 9⅞ inches (25 cm) in total length, but females are usually less than 8⅝ inches (22 cm) long. The tail is very short, barely an inch in length, and lacks the pronounced black tip seen on other weasels. Males usually weigh 1½ to 2⅛ ounces (43 to 62 gm) and females range from 1⅜ to 1⅝ ounces (40 to 45 gm).

These tiny weasels are brown above and white below in summer. In the northern parts of their range they become white in winter. In this region, however, individuals may retain the brown coat into the winter or molt to a partially white coat, retaining some portions of the brown pelage. The least weasel is separated from the long-tailed weasel by its smaller size and in lacking a distinct black tip on its very short tail.

Distribution and Abundance. Least weasels are rare in the upper piedmont and mountains of this region. They have been recorded from Maryland, Virginia, and North Carolina and may inhabit the mountains of northwestern South Carolina as well.

Habitat. Open woodlands, brushy or grassy fields, fencerows, marshes, and

Least weasel (Mustela nivalis).

cultivated fields are the haunts of the least weasel.

Natural History. Least weasels are secretive predators feeding primarily on small rodents, which they hunt throughout their home ranges. They readily investigate any burrow or rockpile in which their slender elongate bodies will fit. They kill by grasping the prey at the base of the neck and then biting through the throat or base of the skull. They do not suck blood as is sometimes rumored, but regularly cache animals for later consumption.

These weasels become sexually mature at 3 to 4 months of age. The least weasel is unusual among mustelids because it reproduces without using delayed implantation. Therefore, reproduction can occur in any season and 2 or 3 litters may be produced

each year. The gestation period is 35 days and the young are born blind and helpless. Their eyes open in about 26 to 29 days, and the young weasels are on their own in 4 to 5 weeks.

Least weasels face most of the same problems as long-tailed weasels and are apparently taken by a number of larger predators. The fur of this weasel has no commercial value.

Long-tailed Weasel
Mustela frenata

Description. This is the most common weasel in the region. The long slender body and neck, short legs, and proportionally long black-tipped tail (40 to 70 percent of the head and body length) are useful identifying features. Upperparts are brown and underparts

Long-tailed weasel (Mustela frenata).

are yellowish white, tending towards whitish in specimens from the piedmont and coastal plain of South Carolina. Although this weasel molts into white winter pelage farther north, the summer and winter colors are similar in this region and only an occasional specimen from Maryland does not remain brown in the winter. Males are much larger than females; adult males weigh 7⅝ to 12 ounces (215 to 368 gm), and females usually weigh 3¼ to 7⅝ ounces (92 to 215 gm). Adult males are 12⅝ to 20 inches (32 to 51 cm) in total length, with a tail almost 4 to 5⅞ inches (10 to 15 cm) long. Adult females are 9⅞ to 16⅛ inches (25 to 41 cm) in total length, and have a tail of 3⅛ to 5½ inches (8 to 14 cm).

Distribution and Abundance. The long-tailed weasel is distributed throughout the four-state region; however, it seldom is seen because of its secretive nature. It probably is more common than suspected in many areas. Population densities fluctuate relative to food availability and environmental conditions.

Habitat. This weasel occupies a wide variety of habitats including woodlands, brushy areas, and borders between woodlands and fields. It is found at all altitudes from mountain peaks to near sea level, and it may exist in close proximity to humans if habitat conditions are appropriate.

Natural History. Long-tailed weasels are highly efficient predators that are abroad throughout the year during both day and night. Their primary food consists of rodents, but they also take moles, shrews, and a variety of other food items including bird eggs, and an occasional cottontail. When natural foods are in short supply, they may take poultry. Their heavy predation on rodents makes them an integral part of their ecosystems.

Long-tailed weasels become sexually mature by 1 year of age. Mating usually occurs during the summer months, but the young are not born until the following spring. This results from a long delay in the development of embryos. Young are usually born in a nest placed in a den in a stump, hollow tree, or some other protected place. Young weasels are born blind and helpless but grow rapidly. The eyes open by 5 weeks of age. Young remain with the mother during their first summer. These weasels are usually solitary except during the mating season.

Snakes, owls, hawks, foxes, and cats occasionally prey on long-tailed weasels. Another mortality factor is man, because this weasel is trapped for its fur and often killed as a threat to poultry. Weasels are important in controlling populations of rats and mice;

for example, in Colorado each weasel was estimated to consume 4 rodents per day. Their great value as efficient mousers should earn this species protection around farmsteads.

Mink
Mustela vison

Description. The mink is characterized by a long thin neck and body, long bushy tail, and short legs. The head is pointed and somewhat flattened above. The eyes and ears are small with the latter almost hidden in the fur. The pelage is generally dark brown above and paler below, usually with a white patch on the throat and chin and other irregular white spots on the belly. The fur of the mink is waterproof, consisting of an outer coat of oily guard hairs and a thick soft underfur. Mink, like all mustelids, have a pair of anal scent glands which produce a very strong musky odor. This liquid may be emitted when the animal is excited, but it cannot be projected in the manner of a skunk.

Males are about 22 to 27 inches (55 to 68 cm) in total length, including a tail of about 7 to 9 inches (18 to 23 cm), and weigh from 2⅝ to 3¼ pounds (1.2 to 1.5 kg). Females are considerably smaller than males. Mink from coastal South Carolina are smaller than individuals from elsewhere in the region.

Distribution and Abundance. Mink are distributed over much of eastern North America and are present throughout this region wherever habi-

Mink (Mustela vison).

tat is suitable. However, they are becoming increasingly uncommon in many areas.

Habitat. Mink are semiaquatic mammals that seldom are found far from water. They are associated with most kinds of wetlands, being at home in

marshes, swamps, and along the borders of lakes, streams, rivers, and even drainage ditches.

Natural History. Mink are generally nocturnal, but occasionally, especially in winter, an individual may be seen abroad during the day. These carnivores eat a variety of both aquatic and terrestrial species, including fish, frogs, crustaceans, birds, and small mammals. They appear to be highly opportunistic, feeding heavily on whatever prey is most available at a particular time.

Mink are usually solitary animals, and males and females come together only briefly to mate in late winter or early spring. The gestation period is about 51 days, and the usual litter consists of 4 young but may range upward to 10. Mink kits are nearly naked at birth and their eyes do not open until about 3 weeks of age; by 6 weeks they begin making short trips from the den. Young mink generally remain with the mother until early fall, using dens that are in hollow logs, under roots of trees, in muskrat burrows, or in the houses of muskrats or beaver.

People are the primary predator of mink. The dense soft underfur makes the pelt valuable to the fur industry and mink are eagerly sought by trappers. Mink coats remain among the most prestigious of fur apparel, and this demand has led to the commercial rearing of millions of these animals in mink ranches throughout the world. The trapping of wild mink is regulated by state wildlife agencies. The primary needs of these animals appear to be the maintenance of adequate unpolluted wetland habitats and close monitoring of population levels

to assure careful regulation of the number of animals trapped.

Eastern Spotted Skunk
Spilogale putorius

Description. The eastern spotted skunk is distinctly smaller than the more familiar striped skunk. The long fur of the spotted skunk has a base color of black, but also has 4 to 6 solid white stripes, which form on the head and neck and break up into a series of irregular white spots along the length of the body. The long bushy black tail is usually tipped with white. Eastern spotted skunks are 16½ to 22 inches (42 to 56 cm) in length, with males being somewhat larger than females. Adult weights range from 2 to 4 pounds (0.9 to 1.8 kg).

Eastern spotted skunk (Spilogale putorius).
Photograph courtesy Great Smoky Mountains National Park.

Distribution and Abundance. This skunk occurs primarily in the highlands from South Carolina to Maryland. In South Carolina there are records from the piedmont in the southwestern portion of the state, but from North Carolina northward the species is restricted to the Appalachian Mountains. It is abundant in this part of its range, and outnumbers the striped skunk in most areas.

Habitat. In mountain habitats, eastern spotted skunks prefer open forests where rocky outcrops provide den sites. They also regularly take up residence around farmyards and under houses and may thus become a nuisance. Generous applications of moth balls usually encourage an unwanted skunk to relocate elsewhere.

Natural History. These small skunks are highly nocturnal, even to the extent of being less active on moonlit nights. They may become inactive during severe winter weather but do not hibernate.

Eastern spotted skunks are omnivorous, eating many insects, other arthropods, and some fruit in summer; however, they also eat birds and many small mammals such as rodents and cottontails. Mammalian food items are especially important in winter diets.

The breeding season in this region begins in late March, and 3 to 6 young are born in a den beneath a house, rock pile, or along a fencerow after an active gestation period of 28 to 31 days. They are born blind and helpless and their eyes do not open for about a month. They are weaned at about 54 days of age.

These skunks are more agile than striped skunks and sometimes climb trees. When disturbed they have the unique behavior of standing on the front feet and arching the tail forward; this handstand posture allows the skunk to see where it aims its spray. The odor of the musk of the eastern spotted skunk is more pungent than that of its striped cousin.

Mortality factors for eastern spotted skunks are similar to those affecting striped skunks, with humans and their automobiles accounting for many deaths. This species may pose a health hazard to humans due to rabies, and care should be taken if individuals become especially aggressive.

Striped Skunk
Mephitis mephitis

Description. This house cat–sized mammal is well known to most people. The long black fur of the body and tail, and the bold white stripe that begins on the head and generally

Striped skunk (Mephitis mephitis).

splits to form 2 parallel stripes down the back, make the striped skunk a very conspicuous animal. The black bushy tail may or may not be tipped with white.

Striped skunks measure about 21 to 27½ inches (53 to 70 cm) in total length, including a tail of about 9 to 13¾ inches (23 to 35 cm); females are as much as 15 percent smaller than males. Adult weight ranges from 2⅝ to 11⅝ pounds (1.2 to 5.3 kg). Individuals from Maryland and northern and eastern Virginia are smaller in size and have much shorter tails than those to the southwest.

This skunk is often detected by its odor. It possesses well-developed scent glands at the base of the tail; these glands are surrounded by special muscles that allow the skunk to propel a charge of the pungent musk for up to 6 yards (5.5 m). The musk,

in addition to its odor, is very irritating to the eyes, and ingestion can cause severe internal difficulties.

Distribution and Abundance. The striped skunk occurs across much of North America and is found throughout the four-state region. Distribu-

tion, however, is somewhat discontinuous, and the species may be absent or very scarce in sizable portions of its range, while being relatively common a few miles away. In general, striped skunks are common in the mountains, uncommon in the piedmont, and rare to absent in the coastal plain.

Habitat. Striped skunks occupy a variety of habitats ranging from high mountain forests to old fields, cultivated lands, and suburban neighborhoods. They generally are associated with upland habitats and are seldom encountered in wetland communities.

Natural History. This skunk is generally nocturnal. In winter it spends the day in a subterranean den, either constructed by the skunk or abandoned by a woodchuck or some other mammal. In summer it may spend the day above ground but usually does not become active until dusk. In more northern parts of its range, striped skunks become inactive in winter.

Striped skunks are omnivorous, with insects making up the primary portion of their diet. They also eat small animals such as mice, frogs, snakes, and bird eggs. Fruits and berries also are consumed.

The breeding season begins in February or March. The gestation period is between 59 and 77 days and litter size ranges from 2 to 10 young; the young are blind and sparsely haired at birth. They open their eyes at about 22 days and are weaned by 8 weeks. Striped skunks may live for 5 to 6 years in the wild and have survived over 10 years in captivity.

Striped skunks are not normally ag-gressive animals and their basic defense is the use, or threat of use, of the musk. When disturbed a skunk faces the intruder, arches its back, elevates its tail, and often clicks its teeth and stamps the ground with its front paws. If the intruder persists, the skunk assumes a U-shaped posture, everts the anus to expose the opening of the scent glands, and discharges the musk toward the intruder. If a skunk becomes very aggressive and persistent in an attack on a human or dog, rabies should be suspected and the skunk reported to local health authorities.

Skunks are preyed upon by a variety of predators such as great horned owls and bobcats. Birds do not appear to be repelled by skunk musk, but they may be affected if the musk gets into their eyes. Skunk pelts were in great demand during the first half of this century, and the animals are still trapped for their fur. To most people, however, the skunk is an animal to be remembered from its strong aroma lingering along the highway or woodland trail.

River Otter
Lutra canadensis

Description. The river otter is a rather large, elongate, semiaquatic animal with a short blunt snout, obvious whiskers, and small eyes and ears. Its neck is thick and its legs are short and stout; the toes are webbed. Its long tail is thick at the base but tapers to the tip. The fur is short and very dense; dorsal color is brown with the

River otter (Lutra canadensis).

sides of the face, chin, and throat usually having a grayish sheen. Adult river otters range from almost 3 to 4 feet (90 to 120 cm) in total length and weigh between 11 and 23 pounds (5 and 10.4 kg). Females are smaller than males, especially in weight.

Distribution and Abundance. The river otter historically occurred along waterways throughout the region. It is still relatively common in portions of the coastal plain from South Carolina to Maryland, but the species appears to be declining in abundance throughout its range and is rare or absent in much of the piedmont and mountains of this region. It is listed as Endangered in Virginia.

Habitat. These mammals are always associated with water. In this region, they are most abundant in coastal estuaries and in the lower reaches of

river systems. They occupy a wide variety of aquatic habitats from streams to lakes where there is a good food supply, clean water, and relatively low levels of human disturbance.

Natural History. Fish are the primary food taken by river otters, but cray-

fish, crabs, amphibians, and other aquatic organisms also are eaten. Otters are apparently not selective when fishing and generally take the most available fish—usually slower species such as suckers, carp, and catfish. In some areas crayfish or crabs become the primary food source. Otters feed heavily between dawn and midmorning, with another peak of activity during the evening. They remain active throughout the year. River otters often are seen in water with only their heads above the surface. Upon diving they leave a trail of bubbles which marks their passage.

The breeding season begins following parturition in late winter and extends into spring. The active gestation period is about 63 days, but there is a long period of delayed implantation prior to the beginning of active pregnancy. Young are usually born in the spring in a den located in a hollow tree, an old muskrat house, or some other protected shelter. Litters usually consist of 2 to 4 young called kits. At birth the kits are blind and helpless but are fully furred. Their eyes open between 21 and 35 days, and the young are weaned by about 3 months of age. They remain with the mother until they are about 1 year old.

River otters are extremely intelligent animals. When encountered along a waterway they exhibit a high level of curiosity. They are reported to engage in extensive bouts of play either by themselves or with other otters.

Adult river otters have few natural enemies. Young individuals may occasionally be taken by predators such as bobcats or alligators, and they are susceptible to both parasites and diseases, but these are not considered important factors in limiting population sizes. The impact of man, however, has been significant. As human activity along waterways increased, levels of water pollution, trapping, and destruction of riparian habitat all have contributed to the decline of this species. There is considerable concern throughout much of the United States about the status of river otter populations, and they may well become endangered in much of this region unless adequate suitable habitat is maintained.

Hair Seals
(Family Phocidae)

Three species of hair seals have been reported from the Carolinas, Virginia, or Maryland—the harbor seal (*Phoca vitulina*), harp seal (*Phoca groenlandica*), and hooded seal (*Cystophora cristata*). All are abundant in the North Atlantic off the coast of Canada and Greenland, and they occasionally wander southward into the coastal waters of this region. Harbor seals are apparently regular visitors, whereas the harp and hooded seals occur this far south only rarely. The walrus occurred as far south as South Carolina, and the gray seal at least to Virginia, during glacial times, but now both are restricted to northern Canada and Alaska. There are also reports of California sea lions from this region, but these surely represent escaped or released individuals.

Description. Hair seals have hind flippers turned permanently to the rear and, therefore, appear very awkward on land. The rather small front flippers are used primarily for steering whereas the large hind flippers provide power for swimming. External ears are lacking and the eyes are large and protruding. The body is smooth and streamlined, and the tail is reduced in size and fits snugly between the hind flippers.

The short fur of the harbor seal is brownish or grayish white with a mottling of darker spots on a somewhat lighter background. Males average about 5 feet 3 inches (1.6 m) in total length whereas the smaller females are usually about 5 feet (1.5 m) long. Males average about 194 pounds (88 kg) and females about 143 pounds (65 kg).

Harp seals are slightly larger than harbor seals, males measuring 5 feet 7 inches to 6 feet 3 inches (1.7 to 1.9 m) in total length and weighing about 298 pounds (135 kg) with females slightly smaller. Adult harp seals are silver-gray in color with a dark harp-shaped saddle on the back; markings on females are less distinct than those on males.

Hooded seals are large, males measuring slightly more than 9 feet (2.8 m) in total length; females may be up to 7 feet 6 inches (2.3 m) long. Males weigh as much as 661 pounds (300 kg) and females weigh about 353 pounds (160 kg) as adults. They are gray with scattered dark spots. The "hood" is an elastic nasal sac which, when inflated, produces a large protuberance between the nostrils and forehead.

Distribution and Abundance. All of these seals are animals of cold northern waters, and all occur infrequently

Harbor seal (Phoca vitulina). *Photograph by Donald F. Kapraun.*

in this region. Harbor seals wander southward almost every year, with most records from late winter or early spring, but there are records for all months except October, November, and December. One harp seal has been observed in this region, near Cape Henry in Virginia in 1945. That individual was captured, photographed, and then escaped. Hooded seals have been recorded but 4 times, twice in North Carolina and twice in Maryland.

Habitat. Harbor seals generally occupy coastal waters, and in this region most records have come from estuaries and river mouths; for example, there are several records from the Chesapeake Bay. They also are seen occasionally on ocean beaches or rock jetties. Harp and hooded seals are animals of the northern ice packs and open ocean. Most southern records are of sick or injured individuals, often found washed up on ocean beaches.

Natural History. Seals are marine predators eating fish, squid, and other marine life. They haul out on remote beaches or ice floes to rest and give birth, but spend most of their lives in the ocean. They are highly adapted to life in cold ocean waters, with a thick layer of insulating blubber beneath the skin and a specialized system of blood vessels in the flippers that helps to regulate body heat. They also have a complex series of physiological adaptations that allow them to remain underwater for as much as 20 to 30 minutes.

Females give birth to a single pup in spring or early summer after a gestation period of nearly a year. The precocial pups enter the water soon after birth. They are weaned after a few weeks and then must fend for themselves. They may live up to 40 years.

Seals face many dangers throughout their lives. Pups are taken by land predators such as polar bears and are killed by man for their fur. In the ocean waters they are attacked by sharks and may come into conflict with humans when they become entangled in fishing nets. Population sizes are often poorly known and much remains to be learned about these northern mammals. They are not likely to become regular components of marine ecosystems in this region, but will continue to provide lucky coastal observers with an occasional glimpse of an intriguing wanderer from the far north.

Mountain Lion
Felis concolor
(Endangered)

Description. The mountain lion is a large, unspotted, long-tailed cat. It is tawny-colored; the sides of the muzzle, the backs of the ears, and the tip of the tail are black. These slender cats are from almost 7 to 9 feet (2.1 to 2.7 m) in total length, up to a third of which is the tail; adults weigh 150 to 200 pounds (68 to 91 kg). Males are 30 to 40 percent larger than females. The kittens are spotted with black patches of fur and have a ringed tail until about 6 months old.

Distribution and Abundance. This magnificent cat once roamed throughout the region. However, it apparently

Mountain lion (Felis concolor).

was eliminated from most of the mid-Atlantic region by the late 1800s. It is now considered by many to be extirpated, although reports of sightings continue in the coastal swamps of the Carolinas and from parts of the southern Appalachian Mountains. The most frequent and best-documented records come from the western counties of Virginia, and it is possible that mountain lions are becoming reestablished in the region.

Habitat. Mountain lions occur primarily in undisturbed habitats which support healthy populations of their primary prey species, white-tailed deer. This cat generally occupies large forested areas and seemingly is at home in coastal swamps as well as on mountain slopes.

Natural History. This shy agile predator is a solitary hunter feeding primarily on deer. In Florida it also eats wild pigs, raccoons, and nine-banded armadillos. The prey is usually stalked and killed by biting through the spinal column in the neck region. After the initial feeding, the carcass of the prey may be cached for a later meal by covering it with leaves and other debris.

Mountain lions begin breeding at about 3 years of age and may mate at any season. After a gestation period of 82 to 98 days, a litter of 1 to 6 kittens (usually 3 in Florida) is born in a den in a hollow log or perhaps on a rock ledge. The eyes are closed at birth but usually open in 8 to 9 days. The young are weaned at 2 to 3 months of age but may remain with the mother into their second year.

Mortality in mountain lions results from many factors. Individuals may be injured or killed during encounters with large prey such as deer, and they are subject to diseases such as rabies.

In spite of having total protection in many parts of its range, the mountain lion continues to be killed by poachers or persons unaware of its status as an Endangered Species in eastern North America; other people are unsympathetic toward large predators. If mountain lions still occur in this region, numbers are very low and total protection is necessary to allow these cats the opportunity to reproduce and to reestablish themselves as a part of our native fauna.

Bobcat
Felis rufus

Description. This is the only short-tailed cat native to the mid-Atlantic region. The color of the bobcat varies from grayish to reddish brown sprin-

kled with small darker spots; the tail is tipped with black only on the dorsal surface. The ears are usually dark and have a white spot near the tip. This medium-sized cat ranges from 24 to almost 40 inches (61 to 100 cm) in total length including a tail of 3½ to 7½ inches (9 to 19 cm). They usually weigh between 10 and 25 pounds (4.5 to 11.3 kg). Males average larger than females in size and weight. Bobcats from Maryland, the piedmont and mountains of Virginia, and the mountains of North Carolina are slightly larger than those from further south.

Distribution and Abundance. The bobcat formerly occurred throughout the middle Atlantic states. It still occurs over most of the Carolinas, but it is found only in the Dismal Swamp and mountains of Virginia and the western half of Maryland. It is relatively com-

Bobcat (Felis rufus).

mon in the coastal plain as far north as the Dismal Swamp and the mountains of the Carolinas and Virginia, but it is uncommon to rare elsewhere.

Habitat. This secretive predator occupies a wide variety of habitats and is present in both coastal swamps and upland forests. It apparently prefers forests where there are extensive areas of dense thickets such as those associated with the early stages of forest succession. In rocky country, ledges and outcrops are important for den sites; bottomlands, pocosins, and swampy thickets provide cover in the coastal plain. Bobcats live in close proximity to people when suitable habitat is available, but their secretive ways often let them go undetected.

Natural History. Bobcats are opportunistic predators taking a variety of

prey species mostly at dusk and dawn. They feed, however, most heavily on rabbits and large rodents such as hispid cotton rats. They also are capable of taking larger prey and are known to occasionally kill white-tailed deer. They depend heavily on sight and hearing to locate their prey and usually stalk to within a few feet of the intended victim and then make a short dash, or pounce, to make a kill.

Bobcats begin breeding at 1 year of age. Mating usually takes place in late winter or early spring. The gestation period is about 62 days, and 2 to 4 kittens are born in late spring or early summer in a den in a rockpile, brushpile, hollow tree, or other similar site. The young are furred but blind at birth. Their eyes open in about 10 days, and by 4 weeks they begin to explore the area around the den. They are weaned by 7 or 8 weeks of age.

The major factor in the mortality of juvenile bobcats appears to be food supply. When food is scarce kittens or juveniles may starve. Adults also may starve in time of low prey availability, and parasites and diseases may weaken both juveniles and adults. Bobcats are also trapped and hunted by man. The price paid for bobcat pelts has risen recently, and in some states biologists have become concerned that increased trapping pressure may threaten this species.

Cetaceans (Whales, Dolphins, and Porpoises)
Orders Mysticeti and Odontoceti

The cetaceans are perhaps the most highly specialized of all mammals. After evolving from land mammals about 60 million years ago, they became fully adapted to life in water. Early biologists classified these animals as fish, but like other mammals they give birth to live young, secrete milk, and, during embryonic development, have a covering of hair. Cetaceans are among the most magnificent and fascinating of all animals by virtue of the great size attained by some, their intelligence, and their unusual ability to exist in an environment foreign to most other mammals. Cetaceans are divided into two distinct groups: the baleen whales in the Order Mysticeti, and the toothed whales, dolphins, and porpoises in the Order Odontoceti.

Cetaceans are found in all oceans and seas. A few, such as river dolphins, inhabit freshwater rivers and lakes. Many species, especially among the odontocetes, occupy the relatively shallow oceanic waters near coastlines. Twenty-nine species of cetaceans have been recorded along the coasts of the Carolinas, Virginia, and Maryland.

The cetacean body tends to be fusiform, except that the snout in several species, such as the sperm and right whales, is blunt rather than beaked or rounded. The head is large and the neck, which is equal in diameter to the trunk of the body, is short and virtually inflexible. The tail terminates in a horizontally flattened fluke which usually is notched. Hind limbs are absent externally, but vestigial girdle and limb bones may persist internally. The front limbs are enclosed in the body to the wrist and each hand is elongated into a paddlelike flipper. The eyes are small, and the ear opening is reduced to the size of pencil lead. The adult body is hairless except for a few bristles that may be on the snout; insulation against cold is achieved with a thick layer of fat, or blubber, beneath the skin.

The skulls of cetaceans are unique among mammals. In most species the bones of the jaws are elongated into a slender rostrum. The nostrils, or blowholes, are positioned high on top of the head, more easily accessible to air at the surface of the water. As a result, many bones of the skull are displaced and overlapped.

Swimming is by vertical movement of the wide, flattened tail. High speeds are attained, enhanced by a streamlined body, smooth skin, and reduced or absent appendages. Whales are remarkable in their ability to remain submerged for long periods of time and to regulate their temperature at great depths, made possible by such physiological adaptations as more efficient lungs, storage of great quantities of oxygen in tissues, and the ability to resist high pressures. Their ability to vocalize and echolocate helps the animals to communicate, locate food, and orient in their aquatic environment. Most are good navigators. The great whales often migrate thousands of miles from polar waters, where they feed in summer, to tropical waters to breed in winter. Less spectacular daily or seasonal movements of many species are correlated with changes in food supply. Cetaceans are thought to be highly intelligent; some whales and dolphins have demonstrated in captivity a remarkable ability to learn and are popular attractions in marine aquaria.

Cetaceans in the Order Odontoceti usually possess teeth, but these vary in number from a single tusk in the narwhal to over 200 in river dolphins. These animals feed on large invertebrates and fish. The skull of toothed whales often is asymmetrical, with a single blowhole positioned to one side of the midline. Those in the Order Mysticeti, however, lack teeth and the roof of the mouth is equipped with numerous baleen plates with which these animals strain small planktonic organisms from the water. The skull of baleen whales is symmetrical, with paired blowholes.

Cetaceans range in size from small dolphins, which measure 6½ feet (2 m) or less and weigh from 50 to 100 pounds (22.5 to 45.5 kg), to the great blue whale which is the largest animal known to have lived, with one individual from the Antarctic being over 98 feet (30 m) in length and weighing about 175 tons (159 MT). Most toothed whales are smaller than baleen whales.

Many products for human consumption are derived from whales, and commercial whaling, extensive in the eighteenth and nineteenth centuries, continues today, primarily by well-equipped fleets of ships from Japan and Russia. The great whales are becoming rare; some species (black right whale, blue whale) may be on the edge of extinction. A strong international conservation and management effort is imperative to assure their survival.

Baleen Whales
(Order Mysticeti)

Seven species of baleen whales have been reported from the Carolinas, Virginia, and Maryland—the minke whale (*Balaenoptera acutorostrata*), sei whale (*B. borealis*), Bryde's whale (*B. edeni*), fin whale (*B. physalus*), blue whale (*B. musculus*), humpback whale (*Megaptera novaeangliae*), and the black right whale (*Balaena glacialis*). The gray whale (*Eschrichtius robustus*), well known for its spectacular migration along the Pacific coast of North America, once lived in the North Atlantic as well; it is omitted here because it has been extirpated from the Atlantic Ocean since colonial times.

Description. Baleen whales are easily identified by the large strips of whalebone, or baleen, that originate from the upper jaw. Most baleen whales also have grooves on the throat and a single ridge on the top of the head extending from the tip of the snout to the blowholes. They are relatively large in size with females averaging slightly larger than males.

The minke whale is the smallest of the baleen whales, attaining a maximum length of 31 feet (9.2 m). It has an extremely narrow and triangular snout, and the ridge on the head is well developed. The tall dorsal fin is located two-thirds back from the tip of the nose. Minke whales are dark gray to black on the back and white on the belly; there is a conspicuous white band across each pectoral flipper. The baleen plates are short and yellowish white, and there are 50 to 70 ventral throat grooves.

The sei whale is blue-gray above and much paler below; frequently there are dark gray or white scars scattered over the body. The baleen is grayish black. Sei whales reach 62 feet (18.6 m) in length and 30 tons (27 MT) in weight. They have a distinct dorsal fin located a third of the total length forward of the tail notch. The snout is less pointed and the ridge on the head, although quite obvious, is less distinct than those of the minke and fin whales. There are 32 to 60 short grooves on the throat.

Similar to the sei whale in external appearance, the Bryde's whale is dark smoky gray on the back, becoming white on the belly. Bryde's whales reach a maximum length of 46 feet (14 m), and have 3 distinct, parallel ridges on top of the head rather than 1 as in most other baleen whales. They possess 40 to 50 relatively long throat grooves, and the strips of baleen are gray in coloration.

The unique coloration pattern of the baleen and the lower lips separate the fin whale from other whalebone whales. The plates of baleen in the front third of the right side are yellowish white, whereas the remaining baleen is striped with alternating bands of blackish gray and yellowish white. The right lower lip is white but the left lower lip is gray. There is frequently a chevron of grayish white skin behind the head on an otherwise grayish black dorsum; the belly is white. The fin whale is large, reaching a maximum length of 80 feet (24 m); the pointed rostrum is narrow and V-shaped, the ridge on the top of the head is well developed, and the head is flat in profile. There are 56 to 100 ventral throat grooves.

The largest animal known to have lived at any time is the blue whale. They reach a maximum length of 87 feet (26 m) and an adult body weight of approximately 100 tons (91 MT) in the northern hemisphere. Other distinguishing characteristics include a reduced dorsal fin that is located a fourth of the total length forward of the tail notch, a broadly rounded rostrum, and bluish color. The baleen is black, and there are 55 to 68 throat grooves.

The long pectoral fins, approximately a third the total body length, and knobby projections on the muzzle and leading edge of the pectoral flippers identify the humpback whale. Humpbacks are heavyset and attain a maximum size of 53 feet (16 m). They are grayish black with irregular patches of white on the belly and pectoral fins. There are 14 to 35 broad throat grooves and the baleen is blackish. The head is broad and rounded when viewed from above; the ridge on the head is indistinct.

The black right whale lacks a dorsal fin and ventral throat grooves, so it is

Humpback whale (Megaptera novaeangliae).
Photograph courtesy Cetacean Research Unit—Provincetown Center for Coastal Studies.

easily identified because all other baleen whales in our area have conspicuous dorsal fins and distinct grooves. Right whales grow to 57 feet (17 m) in length and are heavy bodied, weighing as much as 86 tons (78 MT). They are usually blackish in color on the back and belly, but some individuals have irregular white patches on the underside. The baleen is exceptionally long; usually it is dark but the front strands are sometimes white. Knobby growths, termed callosities, are found on the head; these are irregularly distributed around the 2 blowholes, over the eyes, and on the nose, lips, and chin.

Distribution and Abundance. Minke whales are uncommon in the region. They have stranded in South Carolina, Virginia, and Maryland and also have been sighted off the coast of

North Carolina. The sei whale is rare with single specimens having stranded in North and South Carolina; there is also a sight record of the sei whale from Maryland. Bryde's whales typically inhabit subtropical and tropical waters and therefore are not an important segment of the mammalian fauna of the region; a Virginia specimen constitutes the only record north of Georgia. Fin whales are uncommon in the region; stranding records are available for Maryland, Virginia, and North Carolina. The blue whale is extremely rare and has been recorded only from Maryland. The humpback whale also is relatively uncommon and has been reported only from the Carolinas and Virginia; undoubtedly it occurs off the coast of Maryland as well. The very rare black right whale has been reported from Maryland, North Carolina, and South Carolina;

scientists believe that fewer than 100 right whales exist today along the east coast of North America. The sei, fin, blue, humpback, and black right whales currently are listed by the United States as Endangered Species.

Habitat. The 6 species of baleen whales known to occur regularly in the region spend most of the summer months farther north, and migrate through the coastal waters to and from southern calving grounds in the fall, winter, and spring. The minke, sei, and black right whales pass along the coasts of Maryland, Virginia, and the Carolinas during the colder months of the year, particularly on their northward and more inshore spring migrations. Fin whales move southward and offshore during the fall, and northward and inshore during the spring. The blue whale evidently migrates in the deeper offshore waters; the single stranding record from Maryland occurred in the summer. Humpbacks pass through the shallow coastal waters in the fall and spring on their southward and northward migrations, respectively. Bryde's whales are not known to be highly migratory and presumably spend the entire year in warmer and more southerly waters.

Natural History. Little is known about the natural history of baleen whales because their large size renders them difficult to study and they are infrequently encountered. Most are shallow divers and feed by swimming upward or horizontally with mouth agape, straining krill (a small crustacean), copepods (a form of zooplankton), and small fish through their

baleen. They gain a tremendous amount of weight on their summer feeding grounds and then live off the accumulated blubber as they migrate south during the colder months of the year. Most are solitary (particularly minke and Bryde's whales) or congregate in small groups; during the summer they may form larger feeding pods and work cooperatively to capture food. Pods of 30 to 50 fin whales have been seen feeding together.

Their reproductive patterns are not well known. Species that migrate along the coastline of the region typically give birth to 1 offspring sometime during the colder months, usually between December and March. The gestation period varies from 10 months in the minke whale to between 11 and 12 months in the sei, fin, blue, humpback, and black right whales. The young are weaned at 6 months (minke whale), 7 months (fin and blue whales), 9 months (sei whale), or about 1 year (humpback and black right whales). The Bryde's whale gives birth to a single calf, presumably at any time of the year, but other aspects of its reproductive biology are probably similar to other baleen whales. Adults reproduce every 2 or 3 years, except the minke whale, which likely reproduces every year. Sexual maturity is reached in 6 to 16 years, and they may live as long as 50 years.

Baleen whales have been, and continue to be, taken for their meat, oil, and whalebone, and now several species are on the verge of extinction. Humpback and black right whales were easily taken by early whalers because they migrate very close to shore and are slow swimmers. Advances in the commercial whaling industry such as the harpoon gun and factory ships that process carcasses at sea expedited this depletion, especially of the faster-swimming fin and blue whales. In an effort to protect the remaining whales, the International Whaling Commission determines each year how many of each species can be harvested, but the Commission has no power to enforce these quotas. It should be mentioned, however, that whale hunters have provided scientists the opportunity to gain a tremendous wealth of information concerning the migration, reproduction, behavior, and feeding biology of the great whales.

Rough-toothed, Bottle-nosed, Saddle-backed, Atlantic White-sided, and Risso's Dolphins

These 5 species constitute a rather heterogeneous group but are treated in a single account because many aspects of their biology are similar. The bottle-nosed dolphin (*Tursiops truncatus*) is the best known of the group and its biology is highlighted below; however, information about the rough-toothed dolphin (*Steno bredanensis*), saddle-backed dolphin (*Delphinis delphis*), Atlantic white-sided dolphin (*Lagenorhynchus acutus*), and Risso's dolphin or grampus (*Grampus griseus*) is included when available. Dolphins in this group are distinguished from other species of toothed cetaceans, except other dolphins, by

their relatively small size (less than 13 feet or 4 m), prominent beaks (except Risso's dolphin), and frequently an elaborate pattern of stripes and spots. Means by which they can be separated from other species of dolphins are discussed below.

Description. The rough-toothed dolphin averages 8 feet (2.4 m) in total length and 285 pounds (130 kg) in weight; males are slightly larger than females. The head appears conical because the forehead slopes continually to a long slender beak. The eyes bulge and there are about 94 teeth. A well-developed dorsal fin slants backward at the tip and is located about midway along the back. This dolphin is dark gray above and pinkish white on the belly, throat, and lips; yellowish white scars on the body might be taken for spots, as in spotted dolphins, but these markings result from injuries.

Bottle-nosed dolphins are well known as aquarium attractions and from movies and television programs. This dolphin is relatively nondescript. Large individuals measure about 12 feet (3.7 m) in total length and weigh about 600 pounds (275 kg); males are larger than females. The head tapers quickly to a short but distinct beak, and the dorsal fin is directed backward and centered on the back. Bottle-nosed dolphins are medium to dark gray above and pale gray to whitish on the belly; there are no prominent markings on the body. They have approximately 100 teeth.

The saddle-backed dolphin is vividly colored, with a grayish black back and whitish belly which are separated

Bottle-nosed dolphin (Tursiops truncatus). *Photograph by Peter C. Cram.*

by complicated flank markings; these are yellow or tan on the front half of the body, but pale gray behind. There is a black circle around each eye, a pale gray line extends from the eye to the anus, and another dark gray line extends from the bottom of the lower jaw to the flipper; in contrast, this line runs from the eye to the flipper in striped and spinner dolphins. The saddle-backed dolphin lacks the spots characteristic of spotted dolphins. This small dolphin averages 7 feet (2.1 m) in total length and weighs about 180 pounds (82 kg). The blackish beak is well developed, the dorsal fin is prominent and centered on the back, and there are about 194 teeth.

Another extremely colorful cetacean, the Atlantic white-sided dolphin, is blackish on the back, grayish on the flanks, and whitish on the belly. A distinct patch of white on each side

gradually becomes yellow posterior to the dorsal fin. A gray line extends from the corner of the mouth to the flipper, and a black circle surrounds each eye. This heavyset dolphin averages 8 feet (2.4 m) in total length and weighs about 420 pounds (190 kg); it has about 132 teeth. A prominent and curved fin is located in the middle of the back.

The distinctive body shape of Risso's dolphin immediately distinguishes it from other dolphins; it is most similar to that of pilot whales because it is long and robust, tapers gradually toward the tail, and has no beak. However, the short blunt head of Risso's dolphin has an obvious crease that extends from the blowhole to the upper lip. The tall, pointed dorsal fin is centered on the body. This dolphin is pale gray at birth, but subsequently darkens and then becomes pale with age, especially on the belly and face. Older individuals are heavily scarred. Risso's dolphins average 10 feet (3 m) in total length and 660 pounds (300 kg) in weight; they have about 8 teeth which usually are found only on the lower jaw.

Distribution and Abundance. The rough-toothed dolphin is rare here, and has been reported only from Virginia and North Carolina; apparently this is the northern limit of its range along the west Atlantic Coast. The bottle-nosed dolphin is the most abundant cetacean along the Atlantic Coast, and there are numerous records from the Carolinas, Virginia, and Maryland. The relatively common saddle-backed dolphin has been reported from all 4 states. The Atlantic white-sided dolphin is extremely rare in this region and apparently reaches the southern limit of its distribution in Virginia, where there are 3 recent records. Risso's dolphin is known throughout the region, but is rare.

Habitat. All but the bottle-nosed dolphin are pelagic, occurring in the Gulf Stream or farther east. The saddle-backed dolphin tends to forage in areas where the ocean floor has a substantial amount of topographic relief, such as the edge of the continental shelf. The bottle-nosed dolphin, however, inhabits inshore waters and frequently enters sounds, rivers, and tidal creeks, sometimes ranging many miles inland, such as in the Low Country of southeastern South Carolina or the Chesapeake Bay.

Natural History. These cetaceans feed primarily on squid and fish, but shrimp and octopus also are eaten; Risso's dolphin, in particular, prefers squid. Mortality in these dolphins results from individual and mass strandings, capture in purse seines, and probably predation by sharks and killer whales. Humans seldom hunt them for food or oil, but they are kept captive in aquaria, where they perform with intelligence and agility.

These are gregarious mammals, and a herd can number over a thousand, but herds of 10 to several hundred are most common. The social hierarchy is maintained by echolocation. Individuals of both sexes and all age classes generally occur in a herd.

Mating occurs throughout much of the year, and each female gives birth to a single calf (twins are born occasionally) after a gestation period of 10 to 12 months. Pregnant females sometimes move to the perimeter of the herd to calve. The young are born tail first and then are pushed by the mother to the surface for air. The mother-calf bond is strong. The young nurse for 6 to 18 months. Females apparently reproduce every 2 to 3 years.

Spotted, Striped, and Spinner Dolphins
Stenella species

This very confusing group of dolphins includes, from the Carolinas, Virginia, and Maryland, the bridled spotted dolphin (*Stenella frontalis*), pantropical spotted dolphin (*S. attenuata*), Atlantic spotted dolphin (*S. plagiodon*), striped dolphin (*S. coeruleoalba*), and long-snouted spinner dolphin (*S. longiros-tris*). Another relatively rare species, the short-snouted spinner dolphin (*S. clymene*), has stranded both north (New Jersey) and south (Florida) of the four-state region; its status here is uncertain. Striped and long-snouted spinner dolphins are relatively easy to identify, but the spotted dolphins are virtually impossible to separate because coloration and markings are highly variable. Therefore, we have chosen to limit our discussion of spotted dolphins to the group as a whole.

Description. Dolphins of the genus *Stenella* range in total length from almost 5 to 8½ feet (1.5 to 2.6 m) and weigh as much as 242 pounds (120 kg). Spotted dolphins, as their name implies, have spots except when first born, but the spots vary in number, size, and location by age and from animal to animal. The striped dolphin has a black stripe that extends from the eye to the anus, separating the dark brown or bluish gray back from the white throat and belly; another black stripe runs from the eye to the flipper. The long-snouted spinner dolphin has a distinct tricolored pattern; the back is dark gray, the flanks are pale gray, and the belly and chin are white. However, the greatly elongate and slender snout and the habit of spinning when it breaks the water (hence the common name) are its most diagnostic characteristics. The dorsal fin of spotted and striped dolphins curves backward at the tip, whereas that of the long-snouted spinner dolphin is usually triangular and seldom curved. Also, spotted dolphins are relatively large in size and

have about 148 teeth, the striped dolphin also is relatively large but has about 176 teeth, and the long-snouted spinner dolphin is smaller and has about 210 teeth. Characteristics that distinguish spotted, striped, and spinner dolphins from other dolphins are discussed in the accounts of those species.

Distribution and Abundance. Although there are no specimens from Maryland, the Atlantic spotted dolphin is relatively common throughout this region. The status of the other spotted dolphins, however, is unclear at this time, but apparently North Carolina is the northern limit of the range of the pantropical spotted dolphin; this species is extremely rare and has stranded only once at Cape Hatteras. The striped dolphin is common off the mid-Atlantic states, and has been reported from the Carolinas and Virginia. The long-snouted spinner dolphin is rare in this region, and also reaches the northern limit of its range in North Carolina; it has been reported from South and North Carolina.

Habitat. These are oceanic dolphins, living in tropical and subtropical waters of the Gulf Stream. They are seldom seen within 12 miles (20 km) of the coastline. There is some indication that they migrate inshore during spring and summer, but virtually nothing is known about their migratory patterns in this region.

Natural History. These playful dolphins are highly gregarious and, although groups of less than 100 are encountered most frequently, herds of

Atlantic spotted dolphin (Stenella plagiodon). *Photograph by Howard E. Winn.*

up to several thousand are not un-common. Herds of spotted and spin-ner dolphins contain individuals of all age classes and both sexes, but striped dolphins are segregated by age and sex. These acrobatic and colorful cetaceans frequently are seen leaping from the water or riding bow waves of boats.

Spotted, striped, and spinner dol-phins mate in spring, summer, and autumn. Following a gestation period of almost a year, a single calf is born to each female; twins occasionally are reported. The young usually are weaned in 11 to 12 months. Females reproduce every 2 to 3 years.

Fish, squid, and shrimp form the bulk of the diet. Spotted dolphins for-age near the surface of the ocean, but striped and spinner dolphins are deep-water feeders. They, in turn, are eaten by sharks, killer whales, and false killer whales. Several groups of people in other parts of the world still drive dolphins into shallow water with boats, hunting them for meat and oil. Tuna fishermen also increase mor-tality by trapping them in purse seines.

False Killer Whale, Pilot Whales, Melon-headed Whale, and Killer Whale

This group includes 5 species of cetaceans in the four-state region which can be distinguished from other species of dolphins in the Family Delphinidae by their large size (usu-ally over 13 feet or 4 m), and blunt, beakless heads; all but the melon-headed whale have fewer than 60 teeth. Included in this group are the

Killer whale (Orcinus orca). *Photograph by Walter C. Biggs, Jr.*

false killer whale (*Pseudorca crassidens*), long-finned pilot whale (*Globicephala melaena*), short-finned pilot whale (*G. macrorhynchus*), melon-headed whale (*Peponocephala electra*), and killer whale (*Orcinus orca*). The whales in this group have a well-developed dorsal fin on a predominately black body.

Description. The false killer whale has a long slender blackish body that averages 18 feet (5.4 m) in males and 15 feet (4.7 m) in females; males weigh up to 1½ tons (1.4 MT), and females weigh about half that much. The head is narrow, rounded, and lacks a bulbous forehead; the long, slender, and pointed flippers are located well forward on the body, and have a noticeable "elbow," or bend, in the middle; and the prominent dorsal fin is strongly curved backwards and centered on the back. This whale has approximately 36 teeth.

The bulbous forehead, a low but prominent dorsal fin located in the front half of the long robust body, and the slender pointed flippers distinguish pilot whales from false killer and killer whales. The long-finned pilot whale reaches 20 feet (6.2 m) in total length and weighs up to 3 tons (2.7 MT); females are approximately 15 percent smaller in length and weight than males. Individuals sometimes have pale markings on the throat and belly on an otherwise blackish body. The flipper is long, approximately a fifth the total length, and situated well forward on the body; it has an obvious elbow. There are about 40 teeth. The short-finned pilot whale resembles its long-finned cousin, except its flippers lack an obvious elbow and are only about a sixth of the total length, and there are only 32 or so teeth. This whale is smaller than the long-finned pilot whale, measuring up to 18 feet (5.4 m) in total length in males and 15 feet (4.7 m) in females.

The melon-headed whale resembles the false killer and pilot whales in shape and general appearance; however, it is noticeably smaller, reaching a maximum of 9 feet (2.7 m) in total length and weighing up to 440 pounds (200 kg). The head tapers to a distinct "melon," which is much less pronounced than the bulbous foreheads of pilot whales, the flippers are relatively long and pointed but lack an elbow, and the dorsal fin curves gently backward from the center of the body. The melon-headed whale is uniformly grayish black except for white lips and pale gray patches between the flippers and surrounding the genital area. There are 84 to 100 small, pointed teeth, twice as many as in the other whales described in this account.

The killer whale has a chunky, heavyset body, averaging 26 feet (8 m) in total length and weighing about 8 tons (7.4 MT); females are smaller than males. The blunt rounded head, tall dorsal fin, paddle-shaped flippers, and coloration easily identify this whale. It is black above and white below, and there is an oval white spot above and behind each eye. The dorsal fin in adult males is erect and about 6 feet (1.8 m) high, whereas in females and immature males it is much smaller and curved backwards at the tip; it is centered on the body. There are about 44 teeth.

Distribution and Abundance. False killer whales have been reported from the Carolinas and Maryland, but this relatively rare whale seldom is encountered in the region. The long-finned

pilot whale reaches the southern limit of its distribution in this region; it has been reported from Virginia and North Carolina. Conversely, the short-finned pilot whale reaches the northern limit of its distribution here, and has been recorded from the Carolinas and Virginia. Both species of pilot whales are relatively uncommon. The melon-headed whale is very rare and has stranded only once in Maryland. The killer whale is rare but widely distributed in this region, and has been reported from the Carolinas and Maryland.

Habitat. These whales are primarily oceanic, but pilot whales frequently move inshore when food resources are more plentiful there.

Natural History. These whales are highly gregarious, and pods of several hundred individuals are not uncommon; however, groups of 5 to 50 are encountered most frequently. The hierarchy and cohesiveness of these pods, which includes both sexes and all age classes, are maintained by a complex assortment of vocalizations including a variety of squeaks and clicks.

The false killer whale and both species of pilot whales are well known for mass stranding. There is no consensus as to the cause of this phenomenon, but some of the more plausible explanations relate to becoming trapped in shallow water on a falling tide, lunar or other meteorological conditions, disease and parasitism, fear, acoustical confusion, and social cohesiveness. The killer whale seldom strands en masse.

Squid and fish are consumed in quantity by these whales, and the false killer whale is known to wait for tuna fishermen to release dolphins from purse seines and then catch them. Killer whales also feed on sharks, seals, sea birds, porpoises, sea turtles, and even large baleen whales. Contrary to popular belief, killer whales are not known to eat humans.

Mating occurs year-round in false killer, melon-headed, killer, and perhaps short-finned pilot whales, but in spring and summer in long-finned pilot whales. Each female gives birth to a single young in 12 to 16 months, and the long-finned pilot whale nurses for almost 2 years. The calving interval for these species is about 3 years.

These whales do extremely well in captivity; their intelligence and acrobatic leaps provide entertainment for spectators. They have few natural enemies other than man, who hunts them for meat and oil or out of fear and hostility.

Harbor Porpoise
Phocoena phocoena

Description. The harbor porpoise is the only member of the Family Phocoenidae that enters the coastal waters of the mid-Atlantic region. It is the smallest cetacean in the mid-Atlantic, with adults usually reaching a total length of about 5 feet (1.5 to 1.6 m) and a weight of 100 to 132 pounds (45 to 60 kg). The small rounded head lacks the obvious beak of dolphins, there are no ventral throat grooves, and the tail is notched on the rear edge between the flukes. The back and flippers of the harbor porpoise are dark grayish brown to black and the belly is whitish; there is a narrow stripe that extends on each side of the head from the mouth to the flipper. The triangular dorsal fin is located almost in the middle of the short chunky body. The harbor porpoise has about 92 teeth.

Distribution and Abundance. This porpoise spends summer and autumn farther north in colder subarctic waters, but migrates southward into this region during the winter and spring. Core Banks in Carteret County, North Carolina, is the southernmost point where the harbor porpoise has been recorded along the Atlantic Coast. Yearlings are relatively common in this region from January through May during severe winters, but uncommon to rare during mild ones; adults seldom stray this far south.

Habitat. Inshore waters and shallow coastal bays are the favored haunts of the harbor porpoise. Seldom does it venture into waters deeper than 600 feet (183 m).

Natural History. Mating occurs from June to October. Each female gives birth to a single calf in May, June, or July, and it nurses for approximately 6 months. Thus, females reproduce every other year. Females reach sexual maturity in their second year, males in their third.

The harbor porpoise forages in small groups for fish, shrimp, and cephalopods (octopi and squid). On the other hand, the harbor porpoise is eaten by great white sharks and killer whales. These small porpoises are of little commercial importance, but many aboriginal groups hunt them for meat and oil in other parts of their range. A greater danger resides in their proclivity for shallow water, which often leads them to become entangled in fishing nets or lodged in fish traps. Recent studies have shown the harbor porpoise to have high levels of pesticides and heavy metals in its tissues.

Beaked Whales
(Family Ziphiidae)

Four species of beaked whales have been recorded from the Atlantic waters adjacent to the Carolinas, Virginia, and Maryland. They are the goose-beaked whale (*Ziphius cavirostris*), True's beaked whale (*Mesoplodon mirus*), Gervais' beaked whale (*M. europaeus*), and the dense-beaked whale (*M. densirostris*).

Description. Whales in this family have spindle-shaped bodies that are stocky in the middle and taper at both ends; the snout is long and pointed, and the flukes are not separated by a distinct median notch on the rear edge of the tail. There is a single crescent-shaped blowhole, and the relatively small dorsal fin is placed well back of the midpoint of the body. The lower jaw extends in front of the tip of the upper jaw, and there is a pair of ventral throat grooves that converge under the lower jaw, forming an incompletely closed V.

The body of the goose-beaked whale is long and robust; adults are 18 to 25 feet (5.5 to 7.5 m) in total length and weigh as much as 5½ tons (5 MT). The relatively small head is only about 10 percent of the total length; the beak is less prominent than those of other species of beaked whales. There is a pair of teeth on the tip of the lower jaw in males, but these do not erupt in females. Coloration is highly variable, but, in general, the head is almost always paler than the tail, and juveniles are usually darker than adults. The skin darkens quickly after death, such that an older individual that may be almost completely white while alive becomes grayish black when washed ashore.

True's beaked whale averages about 16 feet (4.9 m) in total length and 1½ tons (1.4 MT) in weight. Its head is small, the beak is pronounced, and males have 2 teeth on the very tip of the lower jaw. True's beaked whale is dark gray to dull black on the back, slate gray on the sides, and white on the belly; pale spots and scratch marks often cover the body. True's beaked whale is much smaller than the goose-beaked whale, and the terminal position of the teeth and the bulging forehead distinguish it from other members of the genus.

The largest member of the genus *Mesoplodon* is Gervais' beaked whale, which reaches a total length of 22 feet (6.7 m) and weighs up to 3 tons (2.7 MT). The body is laterally compressed, much higher than wide, and the small head quickly tapers to a narrow beak. The bluish black skin on the back gradually becomes paler on the belly, and there frequently is an irregular white spot near the genital region. Males also have 2 teeth, but these are positioned 3 to 4 inches (7 to 10 cm) back from the tip of the lower jaw.

The contour of the lower jaw immediately identifies the dense-beaked whale; the upper margin of the lower jaw curves noticeably upward toward the rear corner of the mouth, particularly in males. There also is a single tooth on each side of the mouth in males; it is located on the leading edge of each prominence. Dense-beaked whales are 15 to 16 feet (4.5 to 5.0 m) in total length and slightly over 1 ton (1 MT) in weight. This whale is dark bluish gray above and paler below; grayish white or pink scars occur over much of the body.

Distribution and Abundance. The goose-beaked whale is the most frequently encountered of the beaked whales and has stranded in every state in the region. True's beaked whale has been reported from the Carolinas and Maryland, but Gervais' beaked whale has stranded only in North Carolina. The dense-beaked whale has been reported from Virginia, North Carolina, and South Carolina. These whales are uncommon (goose-beaked whale) or rare (True's, Gervais', and dense-beaked whales), but they probably are present throughout the year in this region.

Habitat. Beaked whales inhabit deep waters off the continental shelf. They are seldom encountered by humans because they infrequently strand and are of scant commercial value. Thus, little is known of their habitat preferences.

Natural History. Beaked whales feed primarily on squid and deepwater fish, but crabs, starfish, and other benthic invertebrates also are eaten on occasion. These whales are not known to have any natural predators; the scarring results from conflicts with other beaked whales, lampreys, cookie-cutter sharks, and parasites.

Small groups of 3 to 10 individuals are typically encountered, although pods with as many as 25 whales have been reported. Some bulls remain solitary except during the breeding season. Apparently, mating is not restricted to any one season and young are born throughout much of the year; in this region, however, pregnant females and juveniles are most often reported in spring and summer.

Sperm Whales
(Family Physeteridae)

Three species of sperm whales inhabit the offshore waters of the Carolinas, Virginia, and Maryland—the sperm whale (*Physeter macrocephalus*), pygmy sperm whale (*Kogia breviceps*), and dwarf sperm whale (*Kogia simus*). Much is known about the biology of the sperm whale because it has been hunted extensively by whalers during this century; in contrast, little is known about pygmy and dwarf sperm whales because they are not commercially important.

Description. The sperm whale can be distinguished from other toothed whales by its tremendous size. Adult males are approximately 50 feet (15.2 m) in total length and weigh about 40 tons (36 MT); adult females average 38 feet (11.6 m) in length and 13 tons

Pod of sperm whales (Physeter macrocephalus). *Photograph by Wayne R. Peterson.*

(12 MT) in weight. The huge head comprises a fourth to a third of the total length and is rectangular in profile; the reduced lower jaw is very narrow, much shorter than the upper, and contains 18 to 25 teeth. The teeth of the upper jaw seldom erupt. A single S-shaped blowhole is located more forward on the head than that of other whales and well to the left of the midline. The pectoral flippers are reduced in size, and a triangular hump, located about a third of the total length forward of the broad and deeply notched tail, replaces the dorsal fin. The sperm whale is grayish brown to bluish black in color except for some irregular white markings around the corners of the mouth and on the belly; the skin is wrinkled in appearance, particularly posterior to the head.

The pygmy and dwarf sperm whales are very similar in appearance and are easily confused. Both are easily separated from other toothed whales by their sharklike body contour, a reduced lower jaw that is much shorter that the upper, and a single blowhole that is situated above the eyes but left of the midline. Pygmy and dwarf sperm whales are dark bluish gray above, becoming paler on the sides and dull whitish on the belly; there is frequently a light-colored line, called a false gill, on both sides of the head behind the eye. The pectoral flippers in both species are located well forward on the body and taper to a blunt point, and the dorsal fin is well developed. Pygmy and dwarf sperm whales differ primarily in size, the pygmy the larger, but adult dwarfs can be as large as juvenile pygmies. In the pygmy sperm whale, however, the dorsal fin is low and located behind the center of the back (high and located near the center of the back in the dwarf sperm whale), there are no teeth in the upper jaw (1 to 3 pairs in the dwarf sperm whale), there are usually 12 to 16 pairs of teeth in the lower jaw (8 to 11 in the dwarf sperm whale), and the teeth are relatively long and thick (short and thin in the dwarf sperm whale). The total length of an adult pygmy sperm whale varies from almost 9 to slightly more than 11 feet (2.7 to 3.4 m), with a weight of 700 to 900 pounds (318 to 408 kg), whereas an adult dwarf sperm whale measures about 7 to 9 feet (2.1 to 2.7 m) and weighs 300 to 600 pounds (136 to 272 kg). The sexes are not appreciably different in size.

Distribution and Abundance. Despite the tremendous number of sperm whales taken by whalers off the coast of the Carolinas in this century, this species remains widespread in the region. It is uncommon, however, and considered Endangered by the U.S. Fish and Wildlife Service. Pygmy and dwarf sperm whales are rare in the region, but this may be a reflection of their preference for deeper offshore waters. The dwarf sperm whale has not been reported from Maryland, but otherwise there are records of all 3 species from the Carolinas, Virginia, and Maryland.

Habitat. Sperm whales favor deeper waters off the continental shelf. They venture into shallow waters to calve or in times of sickness, presumably because they then have a shorter distance to swim to the surface to breathe. They are prodigious divers (the sperm whale may dive to over 3,750 feet or 1,145 m), and can re-

main submerged for over an hour at a time. The shape of pygmy and dwarf sperm whales suggests that they are bottom dwellers.

Natural History. The sperm whale is a gregarious animal, and pods of up to 50 individuals are not uncommon. Mature but reproductively inactive males form bachelor herds, whereas older males are frequently solitary. Several females and their offspring remain together in a nursery school during most of the year, and a large adult male joins the group during the reproductive season. The bull wards off other mature males and mates with receptive females in April and May. A single calf, weighing up to a ton and measuring 11½ to almost 15 feet (3.5 to 4.5 m) in total length, is born after a gestation period of about 15 months. Mothers nurse their young for 1 or 2 years; mature females reproduce every 3 or 4 years. The sperm whale attains sexual maturity at approximately 8 to 10 years of age and may live at least 60 years.

Pygmy and dwarf sperm whales are solitary or congregate in small groups of as many as 10 animals. Females evidently give birth to a single calf every other year sometime between the spring and fall after a gestation period of 9 to 11 months. Offspring remain with their mother for at least a year, and a few mothers and their young form a nursery pod. Calves are weaned at about 1 year of age. Stranding records of pygmy and dwarf sperm whales in the four-state region usually involve females and newborn calves.

It is not known if pygmy and dwarf sperm whales migrate, but the migratory pattern of the sperm whale is relatively well known. Both sexes occur off the coastline of the region year round, and during the fall they begin to move southward to warmer waters. In the spring they begin their return northward migration; in the summer, however, adult males continue to move to colder northern waters, whereas females and the young of both sexes remain in temperate and tropical waters.

Squid and cuttlefish are the primary food of sperm whales, but other prey such as sharks, octopi, and numerous species of bony fish are eaten by the sperm whale, and crabs and shrimp are taken by the pygmy and dwarf sperm whales. The killer whale is probably the only natural enemy of these species.

Man is also an enemy of the sperm whale, having hunted it for its high-quality oil. This oil is found in the spermaceti organ, a large reservoir in the head, and is prized by people in the perfume industry, who use it as a fixative. Another important commodity taken from the sperm whale is ambergris, a musky waxlike substance found in the lower intestine and rectum, and also used as a fixative in the manufacture of perfume. Ambergris is the remains of undigested beaks of squid and octopi. The teeth of the sperm whale are used in scrimshaw, a form of artwork in which drawings are etched on the teeth. The meat of sperm whales is not highly prized; most sperm whales are taken for spermatceti oil, ambergris, and animal food.

Manatees
Order Sirenia

Manatees are familiar in popular literature as the probable basis for the sirens of Greek mythology that seduced mariners with their beauty and melodic songs, causing ships to be wrecked on rocks. In historic times, early European seafarers, obviously too long at sea, told tales of mermaids, lovely creatures that were half woman and half fish and nursed their infants at their breasts. It is assumed that these tales were spawned by sightings of sirenians, which do have a superficial humanoid appearance and paired pectoral mammary glands.

Sirenians are totally aquatic, the only mammals other than whales adapted to spend their entire lives in water. However, they evolved independently of whales, being derived from land-dwelling ancestors that also gave rise to elephants. This relationship is apparent from similarities of their skull and tooth structures.

The Family Trichechidae contains 3 species of manatees in a single genus, which occur along much of the west coast of Africa and along the Atlantic Coast from Virginia to southern Brazil. Only *Trichechus manatus* enters the waters of Virginia and the Carolinas.

Manatee
Trichechus manatus
(Endangered)

Description. The body of a manatee is rounded and fusiform with a small head and a short but somewhat flexible neck. There is a squarish snout with a large, centrally divided, and highly mobile upper lip bearing numerous stiff, vibrissaelike hairs. The remainder of the body is naked except for a few scattered hairs. The eye and ear openings are quite small and there are no external ears. Nostrils are on the upper surface of the snout. Hind limbs are vestigial, not visible externally, and the forelimbs are paddlelike. The tail is an oval fluke, horizontally flattened, and lacks a central cleft. Adults are slate gray to brown and are about 10 to 13 feet (3 to 4 m) in length and weigh up to 1,100 pounds (500 kg). Newborn manatees weigh about 24 to 60 pounds (11 to 27 kg) and are slightly more than 3 feet (1 m) in length.

Distribution and Abundance. This species has been reported in the coastal waters of the Carolinas and southern Virginia as far north as Buckroe Beach, Virginia, where an individual was seen in October 1980. In the United States, its principal stronghold is Florida; its appearance further north apparently is incidental because it cannot tolerate cold water and rarely strays from warmer latitudes.

The manatee presently is consid-

Manatee (Trichechus manatus). *Photograph courtesy U.S. Fish and Wildlife Service.*

ered an Endangered Species by the U.S. Fish and Wildlife Service. It has been hunted for its hide, meat, and oil, but now is fully protected in the United States. Poaching continues, however, and many animals are wounded and killed by propellers of motor boats. While its numbers continue to diminish, progress is being made to develop effective conservation measures.

Habitat. This species lives in coastal waters, estuaries, and freshwater streams bordering tropical and subtropical seas. Neither salinity nor turbidity of the water seems to affect movement, and although it prefers warmer waters, it has been observed in water with a temperature as low as 59°F (15°C).

Natural History. The manatee is strictly herbivorous and will feed on almost any aquatic vegetation, but it prefers grasses. It feeds at night, locating food by touch and smell. The mobile lips close over vegetation and pull it into the mouth, often with the aid of the flexible flippers. This food is chewed by an unending supply of cheek teeth; old worn teeth are pushed forward out of the mouth and replaced from the rear by new unworn teeth.

Though somewhat forbidding in appearance, the manatee is docile, completely harmless, and has few enemies other than man. It usually is solitary, but sometimes gathers in small groups of 2 to 6 animals. The manatee remains relatively dormant on the seabed or river bottom during the day and can remain submerged up to 20 minutes, but it normally surfaces to breathe every 2 to 5 minutes.

Mating occurs throughout the year and is preceded by courtship of a female by numerous bulls. Length of the gestation period is probably about 385 days; usually a single calf is born. Suckling occurs underwater with the cow in a horizontal position; she may nurse her young for as long as 2 years. Sexual maturity is attained at the age of 4 to 6 years.

Hoofed Mammals
Order Artiodactyla

There are 2 orders of hoofed mammals: the Perissodactyla, or odd-toed ungulates, such as horses, asses, zebras, tapirs, and rhinoceroses; and the Artiodactyla, or even-toed ungulates, such as pigs, peccaries, camels, giraffes, hippopotami, pronghorn, deer, and bovids (including bison, cattle, sheep, goats, and antelopes). Hoofed mammals are found on all continents except Australia and Antarctica and are highly diverse; however, few non-domesticated ungulates occur in this region, and they are all in the order Artiodactyla.

Ungulates are most distinct from other mammals in the specializations of their limbs. They walk or run on the tips of their toes, with the sole and heel of the foot raised off the ground. The end of each digit is flattened and protected by a hoof. The limbs are elongate, and, unlike those of other mammals, the lower segment of each leg is longer than the upper segment. The bones of the foot are fused, in varying degrees in different species, to form cannon bones. The weight of the animal is borne on the ends of the digits, the cannon bones, and the bones of the ankles and lower legs. The "cloven-hoofed" Artiodactyls have either 2 or 4 digits per foot. The third and fourth digits are always present and about equal in size; the second and fifth digits, if present, usually are small or are functionless dewclaws, as in the deer and bovids.

Artiodactyls are herbivores. Specializations of teeth and digestive system relate to their diet, which consists of a variety of plant material that is difficult to chew and digest. The upper incisors are absent in such species as deer and cattle; they crop grasses and foliage by means of the lower incisors biting against hardened tissues in the roof of the mouth, and with the aid of a large, mobile, and protrusible tongue. Upper canines in most species also are absent or greatly reduced. The cheek teeth are the principal teeth for chewing. They are broad and flat with strong cusps or crescents and tend to be high crowned to allow for wear. They are effective in crushing and grinding tough plant tissue.

The stomach of some artiodactyls is complex, consisting of 2 or more compartments. Plant tissue is relatively low in nutrients and includes a great quantity of cellulose for which mammals lack digestive enzymes. However, cellulose-digesting bacteria and protozoa which reside in the stomach make it possible for ungulates to derive nutrients from this food source. Vegetation is first swallowed into the rumen of the stomach and is partially broken down by the action of microorganisms. Later, the cud is returned to the mouth, chewed again, reswallowed, and passed progressively to other stomach compartments where digestion proceeds. Cud-chewing allows an animal to ingest quickly large quantities of vegetation and later to chew and digest it at leisure, safe from predators.

Ungulates are the only modern mammals to possess horns and antlers. True horns, found only among the bovids, are unbranched and permanent. They consist of an inner core of bone, an outgrowth of the skull, and an outer sheath of hardened epidermal, or skin, tissue. Horns grow throughout the adult life of the animal. In many species they are present on both sexes, but usually are larger on the males.

Antlers occur as head ornaments on males of various species of deer, and also on female reindeer and caribou. They are formed entirely of bone and are branched. A growing antler is covered by a layer of skin, or velvet, which dies and is stripped away when the antler is fully grown. The antler is shed annually after the mating season; a new set begins to grow each spring. These structures are employed primarily for ritualized displays and fighting during the rut or breeding period.

Representatives of 3 families of the Order Artiodactyla either occur within the Carolinas, Virginia, and Maryland at present or did so in historic times.

Family Suidae includes feral pigs (of domestic origin, living in the wild), as well as the introduced European wild boar. They are characterized by the presence of large ever-growing canine teeth, forming tusks that protrude from the mouth. Limbs are short and the foot structure is somewhat different from that of other members of the order.

Family Cervidae includes the deer, characterized by bony antlers, no upper incisors, and usually no canines. The white-tailed deer is the only native member of the order surviving in the region. Elk were extirpated in the late eighteenth century. Sika deer, never indigenous to the region, were introduced from the Orient and now exist in small populations on the Eastern Shore of Maryland and Virginia. The fallow deer, a native to Europe and parts of west Asia, also was introduced into Maryland and North Carolina, but apparently there are no established herds today.

Family Bovidae includes the bison, extirpated from this region. It is a grazing animal characterized by the presence of permanent horns consisting of a bony core and a keratinized outer sheath, usually present in both sexes.

Wild Pig
Sus scrofa
(Introduced)

Description. Wild pigs in this region are of 2 origins: some are descendants of domestic hogs that have become feral, whereas others are the descendants of wild pigs (usually referred to as the European or Russian boar) imported into hunting preserves in western North Carolina. Both belong to the same species and some interbreeding has occurred where they have occupied the same ranges. Typically these wild pigs are similar to domestic hogs but are thinner and have coarser coats of fur; their tusks (elon-

Wild pig (Sus scrofa).

gate canine teeth) are generally longer and sharper than those seen in domestic stock. The European or Russian hogs of the North Carolina mountains tend to be heavier in the forequarters than domestic hogs and may develop a mane of long bristles down the back; each of these long hairs usually is split at the tip. These hogs are usually black, but the tips of the long guard hairs (bristles) are silvery gray or brown. They also have an undercoat of finer hair. Their tusks are usually 2 to 4 inches (5 to 10 cm) in length, and in exceptional cases they may be almost 8 inches (20 cm) long. Wild pigs stand as much as 3 feet (0.9 m) at the shoulder and weigh as much as 400 pounds (181 kg).

Distribution and Abundance. Feral hogs are relatively common in portions of the coastal plain of South Carolina and in isolated pockets elsewhere in the region. The European or Russian boar was originally stocked in a hunting preserve on Hooper's Bald in Graham County, North Carolina, in 1912. It subsequently was stocked at several other locations in the Appalachian Mountains. Escaped and released individuals spread and prospered, and wild pigs are now fairly common in the Great Smoky Mountains.

Habitat. In the eastern portion of the region, wild pigs are most often associated with bottomland hardwood forests along coastal plain rivers. Those of the mountains generally inhabit lower elevation forests in winter and move to higher elevation hardwood forests in summer.

Natural History. Wild pigs are omnivores, eating a wide variety of foods.

They dig for tubers and roots or forage on the shoots of herbaceous plants. They also feed on many fruits, and during fall and winter depend heavily on acorns and other mast. They generally become active late in the evening and do much of their feeding at night.

Maturity is reached prior to 1 year of age. Two peaks of reproductive activity appear to occur in the wild pig population in the North Carolina mountains—the first in early summer and another in midwinter. The gestation period is apparently similar to domestic hogs at 112 to 114 days. A litter comprises an average of 4 to 5 dappled young. The piglets are usually born in shallow beds that may or may not be lined with vegetation. The young pigs follow the mother for short distances soon after birth, and by 3 weeks of age they accompany her on foraging trips.

Predators such as bobcats and foxes may take young pigs, and black bears occasionally prey on juveniles and adults. Wild pigs also are hunted by man. Feral hogs are generally not covered by game laws, and in fact may be considered a nuisance. The European boar of the mountains is considered a game animal in North Carolina and thus is protected by closed seasons and bag limits. It is, however, considered a threat to natural ecosystems in the Great Smoky Mountains National Park, and there is an active program to remove hogs from the park. These opportunistic mammals often compete with native animals for food and damage forest ecosystems while rooting for food, resulting in a significant reduction in the numbers of individuals of several species of herbaceous plants in the high altitude hardwood forests. Wild pigs are thus an interesting but controversial part of our mammalian fauna.

Wapiti or Elk
Cervus elaphus
(Extirpated)

Description. The elk is a large deer with a dark reddish brown summer coat. Females remain dark throughout the year, but in winter the bulls are pale cream with a darker mane. In both sexes there is a yellowish rump patch. Males have large antlers which may extend almost 5 feet (1.5 m) above the head. Elk, in contrast to other species of North American deer, have 2 upper canine teeth. Bull elk are almost 8 feet (2.4 m) in total length and stand about 4½ feet (1.4 m) at the shoulder. A large bull may weigh 1,100 pounds (499 kg) but males average about 650 pounds (295 kg). Females are about 20 to 25 percent smaller than males.

Distribution and Abundance. Elk were widespread in North America when the first European settlers arrived, but by the mid-1800s the species had been extirpated from this region. They were, however, reintroduced into several Virginia counties in 1917 and again in 1935, and into North Carolina in 1912 and again in the 1940s. They prospered for several years in Virginia but then became infected with a lethal parasite from the white-tailed deer. The last elk seen

Elk (Cervus elaphus). *Photograph by Walter C. Biggs, Jr.*

in Virginia was in the Giles-Bland Mountain Range In 1974; those animals introduced into North Carolina never became established.

Habitat. Little is known about the habitats occupied by elk in eastern North America. In the western states they occupy mountain woodlands where there are open meadows; they migrate to lower elevations in winter. Similar open forests with adjacent meadows or fields may have provided suitable habitat in the east.

Natural History. Elk eat a variety of foods including shrubs, forbs, and grasses. Their diet varies throughout the year as they utilize those foods most abundant and readily available.

Mating occurs in autumn, with the bulls becoming very vocal during the rut. This species is polygamous and males gather a harem of females at this time. The gestation period is about 8½ months, and a single precocial calf is born usually in early summer; it may not be fully weaned for 9 months.

Elk face many of the same problems as other species of deer. In colonial times they apparently retreated before the advance of settlers and disappeared first from the more settled areas and last from the more remote mountain regions. The reestablished herd in Virginia was legally hunted and likely further reduced by poachers and by farmers concerned about crop depredations. Extirpation of this recent population apparently was due, however, not to direct killing or even loss of habitat, but to the introduction of a parasite against which it had no immunity.

Sika Deer

Cervus nippon
(Introduced)

Description. This small deer is easily separated from the native white-tail by the presence of white spots on the adults and by the reddish color of the pelage. It also has a dark line down the middle of the back and a distinct white rump patch; the underside of the tail has less white than that of the white-tailed deer. The sika deer has erect antlers with the individual tines branching from the front of the main beam. It is about 4 to 5 feet (1.2 to 1.5 m) in total length and stands about 2½ feet (0.7 to 0.8 m) at the shoulder. Males are much larger than females.

Distribution and Abundance. A native of China and Japan, the sika deer has been introduced into several countries including the United States. It was introduced onto the Eastern Shore of Virginia and Maryland in the early 1900s and has prospered on Chincoteague Island in Virginia and on Assateague Island in Virginia and southern Maryland. It also occurs on the mainland in Dorchester County, Maryland, and perhaps adjacent to Chincoteague Island in northeastern Virginia. These deer may be readily observed on the Chincoteague National Wildlife Refuge.

Habitat. The sika deer often feeds in the shallow brackish marshes at Chincoteague, but it may also be encountered in the loblolly pine forests and dense shrub thickets common on Chincoteague and Assateague islands.

Natural History. These are primarily nocturnal foragers, but they often are

Sika deer (Cervus nippon) *in summer pelage.*

Sika deer (Cervus nippon), *female in winter pelage.*

seen grazing in the open marsh in early morning or late afternoon. They appear to browse a variety of plant species such as greenbrier and poison ivy and also graze the herbaceous plants of the marsh and forest floor.

Males establish territories during summer and engage in fights with other bucks during the rut in fall. These deer are polygamous and each male gathers several females into his territory during the mating season. The gestation period is about 7 months, and a single calf is born, usually in late spring.

Sika deer may be hunted legally in both Maryland and Virginia, and likely face most of the same problems as the native white-tailed deer. They appear, however, to be doing well on Assateague where numbers have been increasing in recent years. Carefully regulated hunts have been initiated on the Chincoteague National Wildlife Refuge to help control the size of the herd.

White-tailed Deer

Odocoileus virginianus

Description. This is the only native deer present in the middle Atlantic region. It is identified easily by the completely white underside of the tail, which is usually elevated when the deer takes flight. White-tailed deer are generally reddish brown in color in summer but molt and take on a grayish brown color in winter. Fawns are spotted with white until their first molt in the fall of their first year. There is a white patch on the throat and a white band across the nose; underparts are also white. Albino deer occur occasionally, as do individuals with a mosaic of reddish brown and white patches of fur.

Male white-tailed deer usually have conspicuous antlers. Three forms are recognized: straight peglike structures called spikes, longer unbranched antlers often called cowhorns, and branched antlers, each consisting of several points or tines. In eastern North America all points are usually counted when antlers are evaluated; a 10-point buck has 5 tines on each antler. In western states, however, only a single antler is counted, and so a 10-point buck in Virginia becomes a 5-point buck in Colorado. Each branched antler of the white-tailed deer normally consists of a main stem from which each individual tine branches. The first point, or brow tine, branches to the inside, but all others branch outward from the main stem. Males grow their first set of antlers during the summer that they become a year old. Often the first set consists only of spikes, but in subsequent years the size of the antlers and the number of points is controlled

White-tailed deer (Odocoileus virginianus), *buck in winter pelage.*
Photograph by Walter C. Biggs, Jr.

more by genetics and the general health of the deer than by age. Females occasionally grow antlers, but these are very uncommon and usually small. White-tailed bucks average about 6 feet (1.8 m) in total length and slightly more than 3 feet (1.0 m) at the shoulder. Weight of adults is usually between 50 and 350 pounds (23 to 159 kg) and averages about 125 pounds (57 kg) in males. Females are somewhat smaller than males. White-tailed deer from most coastal islands are smaller than those from the mainland.

Distribution and Abundance. The white-tailed deer now occurs in most counties in the four-state region, although it was eliminated from many of the more heavily settled areas in the early 1900s. Restocking efforts, active management programs, and changes in agricultural practices have brought about a dramatic increase in the number of deer. There are probably more white-tailed deer in this region now than ever before. In fact, in many areas they are so abundant as to become a threat to destroy crops and to overbrowse the natural vegetation.

Habitat. The white-tailed deer is at home in most of the natural communities of the region and may be encountered in coastal marshes as well as high mountain forests. Prime habitat, however, appears to be broken areas of mixed young forests, old fields, and crop lands typical of much of the rural portions of the region.

Natural History. Deer are primarily browsers, feeding on the leaves and twigs of a wide variety of plants. However, they are quite selective when choices are available, and show an as-

White-tailed deer (Odocoileus virginianus), *molting doe.*

White-tailed deer (Odocoileus virginianus), *fawns.*

tonishing ability to select the most nutritious foods from among those present. In addition to browsing on leaves and twigs, deer eat acorns and other mast as well as a wide variety of herbaceous plants. White-tailed deer also make heavy use of agricultural crops such as corn and soybeans. This deer feeds primarily at dawn and dusk, but it often feeds at night and during daylight hours as well.

White-tailed deer may breed during their first year, and most become reproductive by the second year. Mating usually occurs in autumn, and during this rutting period bucks become very aggressive and are more active than usual during daylight hours. The gestation period is about 201 days, and 1 or 2 precocial young are born in the spring. Although the white-spotted fawns are capable of coordinated movement very early, they are often left in a hidden spot by the mother while she feeds. Such hidden fawns usually are assumed to be lost when discovered by humans, but should not be disturbed. Fawns are weaned at about 8 months of age and may remain with the doe for over a year.

White-tailed deer face many hardships. They are host to a variety of parasites, most noticeably the giant liver fluke. They may also contract a variety of diseases that cause weakness or death. The greatest cause of mortality, however, is man. Deer are hunted legally in all states within this region, and it is likely that the illegal kill by poachers is equal to or greater than the legal take in some areas. Many deer also are killed or crippled by collisions with automobiles. In places where deer are doing significant damage to agricultural crops, legal or illegal removal of deer by farmers also may be significant. In spite of heavy mortality, white-tailed deer are increasing in number in many areas.

Bison

Bison bison

(Extirpated)

Description. The bison, or buffalo, is usually thought of as a mammal of the western grasslands. When European settlers first arrived in the Appalachian Mountains, however, they found bison present. The large head and massive forequarters give bison a hump-backed appearance; this coupled with upturned horns and a shaggy beard from chin to throat makes them very distinctive. They are large mammals, with males weighing as much as 1,800 pounds (818 kg) and females weighing up to 1,200 pounds (545 kg). Bulls may reach almost 6 feet (1.8

m) at the shoulder, and are thus the most massive native terrestrial mammal of North America. Females are somewhat smaller, reaching almost 5 feet (1.5 m) at the shoulder.

Distribution and Abundance. Bison once ranged over the western parts of the Carolinas, Virginia, and Maryland. They were the first native mammal to be extirpated by man and were gone from North Carolina by 1765, Maryland by 1775, and Virginia by 1797. A small group was unsuccessfully reintroduced into western North Carolina in 1912. Bison are now maintained in captivity at several locations but do not occur in the wild in eastern North America.

Habitat. The eastern bison was apparently better adapted to woodlands than was the western plains form. Early records indicate that eastern bison occurred in small herds and favored the open valleys of the piedmont and mountains.

Natural History. Very little is known about the biology of the eastern bison, but it was presumably similar to the plains-dwelling bison of the west. These are gregarious grazing animals foraging on a variety of grasses, sedges, and broad-leafed plants. Like cattle, bison are cud chewers and have a four-chambered stomach.

Mating occurs during the summer, and after a 9 to 9½ month gestation period usually a single calf is born. Sexual maturity is reached in 2 to 3 years.

Adult bison have few enemies other than man, who eliminated the bison from the eastern United States and

Bison (Bison bison). *Photograph by Walter C. Biggs, Jr.*

very nearly eliminated it from the western plains as well. Small numbers of bison remain in the west, the result of one of the first successful conservation efforts in this country. It has not, however, been returned to eastern habitats except in small numbers under captive conditions.

Glossary

Altricial: Young that at birth require prolonged parental care, usually born blind, without hair, and unable to walk (as opposed to precocial).

Antler: A branched, bony ornament on the head of deer which is shed annually and covered with soft spongy skin, often referred to as velvet, when growing.

Arboreal: Living in trees.

Arthropod: Member of Phylum Arthropoda; examples are insects, spiders, and crustaceans such as crayfish.

Bald: Natural grassland occurring on high mountain peaks.

Blowhole: Nostril on top of skull in whales that serves as a passageway for air and water vapor when the animal surfaces.

Blubber: Layer of fat beneath the skin of cetaceans.

Boreal forest: Coniferous forests of the northern United States and Canada.

Bovid: Ungulate or hoofed mammals in the Family Bovidae, such as cattle and sheep.

Cache: To conceal or store; hide in a secret place.

Calcar: A cartilaginous spur from the heel of bats which helps support tail or interfemoral membrane.

Canine teeth: Usually long and pointed, adapted for grasping and piercing; located between the incisors and cheek teeth.

Cannon bone: Bone formed in ungulate mammals by fusion of bones of the foot that helps support the weight of the animal.

Carnivorous: Meat-eating.

Cetacean: Members of the mammalian orders Mysticeti and Odontoceti, including all whales, porpoises, and dolphins.

Cheek teeth: Teeth behind the canines; premolars and/or molars.

Coastal plain: The emergent portion of the continental shelf inland to the fall line.

Commensal: Living with another in a close, but nonparasitic, relationship.

Community: A group of interacting organisms of various species living together in a common environment.

Coniferous: Cone-bearing trees, as pines, firs, hemlocks, and spruce.

Cursorial: Locomotion involving running.

Deciduous: In plants, shedding or losing leaves at the end of a growing season (as opposed to evergreen).

Delayed fertilization: A phenomenon occurring after copulation when fertilization of eggs is delayed for a period of time; viable sperm are retained in the reproductive tract of the female. This occurs in many bats.

Delayed implantation: A phenomenon occurring after fertilization when implantation of embryo(s) into wall of uterus is delayed, thereby arresting development for a period of time before normal gestation. This occurs in some mustelids.

Diastema: A gap between teeth in a jaw, often between incisors and the cheek teeth due to absence of the canines.

Digitigrade: Walking on the toes with the bones of the wrist and ankle raised off the ground, as in dogs and cats.

Diurnal: Pertaining to daylight hours; to be active during the day (as opposed to nocturnal).

Dorsal: Pertaining to the back or upper surface of an animal (as opposed to ventral).

Echolocation: Use of emitted high frequency pulses to locate an object by interpreting echos reflected from that object; present in most bats and some cetaceans and shrews.

Ecologist: One who studies the relationships between living organisms and their environment.

Endangered species: A species of living organisms on the verge of becoming extinct; when capitalized it refers to a species officially so designated by the U.S. Congress or a state agency and provided special protection under federal or state law.

Evergreen: Plant having foliage that remains green until new foliage is formed (as opposed to deciduous).

Extinct: No longer living, all individuals of a species having died.

Extirpated: No longer present in a portion of a species' original range of distribution.

Fall line: Boundary between the piedmont and coastal plain, marked by falls or rapids along streams flowing eastward.

Feral: An animal, formerly domesticated, which has reverted to living in the wild.

Flipper: A vertebrate appendage adapted for aquatic locomotion, with digits fully webbed, as in seals, cetaceans, and manatees.

Floodplain: An area bordering a river, subject to periodic flooding.

Fluke:
(1) Lateral flattened lobes of the tail of cetaceans and manatees, supported by dense fibrous tissue.
(2) A parasitic flatworm.

Forage: To search for food.

Forb: Broadleaf herbaceous plants other than grass.

Fossil: Remains of organisms, preserved in layers of sediment in the earth's crust, providing a record of the past history of life.

Fossorial: Adapted for digging; pertaining to life under the surface of the ground.

Fusiform: Tapering toward the ends; streamlined.

Genus: A group of related species; similar genera are grouped in a family.

Gestation period: Time required for the development of young from fertilization to birth; the period of pregnancy.

Guard hairs: The outer coat of coarse, longer hairs that cover and protect the underfur; found on most mammals.

Habitat: Place where an organism normally lives.

Herbivorous: Feeding primarily on vegetation; plant-eating.

Hibernaculum: Shelter or place where an animal hibernates.

Hibernate: To pass the winter in a state of torpor or dormancy.

Home range: The area occupied by a mammal during its life, where its normal daily activities are carried out.

Horn: Structure projecting from the head of such bovid mammals as cattle and sheep used for offense, defense, or social display. It is formed of a bony core covered over by a permanent hollow sheath of hardened and thickened tissue.

Hummock: A small knoll or hillock rising above the surrounding terrain.

Hybrid: Offspring that result from interbreeding between individuals of two species, such as wolf and coyote.

Incisor teeth: Front teeth; anteriormost of the four basic types of teeth of mammals, adapted for gnawing or shearing.

Insectivorous: Feeding on insects.

Interfemoral membrane: In bats, the integumentary membrane stretching between the hind legs, including all or a part of the tail.

Lichen: An organism often growing on rocks, tree trunks, or the soil, which is a combination of fungi and algae. It is used as food by some mammals.

Litter:
(1) The group of young produced by a female at one birth.
(2) Accumulation of such organic matter as leaves, etc., on the surface of the ground.

Marsupium: The external pouch on the abdomen of marsupial mammals. It encloses the mammary glands and serves as an incubation chamber.

Mast: Nutlike fruit of trees such as oak, hickory, beech, etc., used as food by many mammals, including squirrels.

Maternity colony: A group of nursing mammals and their young.

Melanistic: Having an excess quantity of dark pigment.

Migration: Regular movement of a species from one part of its range to another, often on a seasonal basis.

Molt: The shedding and replacement of hair.

Mustelids: Members of the Family Mustelidae (Order Carnivora), which includes weasels, skunks, mink, and others.

Nictitating membrane: A third, usually transparent eyelid present in some species of mammals which can be moved over the surface of the eye.

Nocturnal: Pertaining to nighttime; active during the hours of darkness (as opposed to diurnal).

Omnivorous: Having a varied diet; feeding on both plant and animal matter.

Parturition: The process of giving birth.

Pelage: All the hair covering the body of a mammal.

Pelagic: Pertaining to the open ocean or to an animal that lives at sea.

Piedmont: An area of low rolling topography between the fall line and the mountains.

Placenta: The vascular structure, produced jointly from embryonic and maternal tissues, through which embryos are nourished while in the uterus of the mother.

Plankton: Passively floating animal and plant organisms in a body of water; food supply for some marine mammals.

Plantigrade: Walking or standing with sole and heel of foot touching the ground, as in bears and humans.

Pocosin: An evergreen shrub bog or low forest in the coastal plain of the Carolinas, characterized by highly organic soils and long periods of flooding.

Polygamous: Having more than one mate at a time.

Precocial: Young that at birth require little parental care. They usually have sight and fur and are capable of moving about (as opposed to altricial).

Predator: A species that preys upon other animals for its food.

Prehensile: A structure, as the tail of an animal, adapted for grasping or holding, as by coiling around.

Prey: An animal seized by another for food.

Riparian: In or on the bank of a natural watercourse such as a river.

Rumen: The first compartment of the complex stomach of some ungulate mammals, such as cattle.

Runway: A worn and detectable pathway resulting from repeated usage by some mammals; often produced by rodents.

Rut: An annually recurring state of sexual excitement, as in deer; the time when breeding occurs.

Saltatorial: Movement by leaping.

Sedentary: Settled in one place; inactive; sluggish.

Sign: Any evidence of the presence of an animal, e.g., trails, footprints, scat, etc.

Social order: A system of ranking within a group of animals; determines precedence and helps avoid conflict.

Species: The basic unit used in classifying animals; a population or a group of populations of closely related and similar organisms that are capable of interbreeding in their natural environment to produce fertile offspring.

Subspecies: A genetically and geo-graphically distinct race or subdivision of a species.

Succession: The orderly sequence of communities that develop over time after removal of existing communities, such as old-field succession following the abandonment of agricultural lands.

Succulent plant: One that is juicy and with soft tissues.

Talus slope: The sloping mass of rock fragments below a cliff.

Territory: An area defended by an animal or group of animals from other individuals or groups of the same species.

Tine: A prong or spike on the antler of a deer.

Torpor: A state of dormancy or inactivity in many mammals, accompanied by lowered metabolism, body temperature, and heart and breathing rates.

Tragus: A fleshy projection from the lower margin of the ear in most bats; important in echolocation.

Underfur: Coat of dense fine hair, usually overlain by guard hairs, serving primarily for insulation.

Ungulate: A hoofed mammal.

Unguligrade: Walking with only the tips of the toes in contact with the surface; in hoofed mammals.

Uterus: An organ of the reproductive system of female mammals within which embryos develop.

Vertebrate: An animal in Subphylum Vertebrata, having a vertebral column as the main axial support of the body; includes fishes, amphibians, reptiles, birds, and mammals.

Vestigial: A remnant of a structure which once was of greater importance.

Vibrissae: Long, stiff hairs on the snout of a mammal serving primarily as tactile receptors; whiskers.

Wean: To transfer dependence of young mammals from mother's milk to other sources of nourishment.

Wetland: An area seasonally saturated with water to the extent that most plant species present are adapted to life in wet soils; usually marshes and swamps.

Selected References

Regional Publications

Bailey, J. W. 1946. *The Mammals of Virginia*. Williams Printing Co., Richmond, VA. A description of the mammals known from Virginia and many aspects of their biology.

Bailey, V. 1923. Mammals of the District of Columbia. *Proceedings of the Biological Society of Washington* 36: 103–38. A list of mammal species with notes on their biology.

Feldhamer, G. A., J. E. Gates, and J. A. Chapman. 1984. Rare, Threatened, Endangered and Extirpated Mammals from Maryland. In *Threatened and Endangered Plants and Animals of Maryland*, edited by A. W. Norden, D. C. Forester, and G. H. Fenwick, 395–438. Maryland Natural Heritage Program, Special Publication 84-1. A list of mammals from Maryland, including information on endangered, threatened, and poorly known species.

Golley, F. B. 1966. *South Carolina Mammals*. Contributions of The Charleston Museum, Number 15, Charleston, SC. An introduction to mammalogy in South Carolina, information on the study of mammals, and a description of the mammals of the state including keys to identification and distribution maps.

Hamilton, W. J., Jr., and J. O. Whitaker, Jr. 1977. *Mammals of the Eastern United States*. 2d edition. Cornell University Press, Ithaca, NY. Accounts of all species of wild mammals known to occur in eastern North America, including keys to similar species and distribution maps.

Handley, C. O., Jr. 1971. Appalachian Mammalian Geography—Recent Epoch. In *The Distributional History of the Biota of the Southern Appalachians. Part III. Vertebrates*, edited by P. C. Holt, R. A. Paterson, and J. P. Hubbard, 263–303. Virginia Polytechnic Institute, Blacksburg, VA. A discussion of the effect of past and present climate on current mammal distributions in the Appalachian Mountains.

———. 1979. Mammals of the Dismal Swamp: A Historical Account. In *The Great Dismal Swamp*, edited by P. C. Holt, 297–357. University Press of Virginia, Charlottesville, VA. Current status and history of mammals of the Dismal Swamp region.

———. 1980. Mammals. In *Endangered and Threatened Plants and Animals of Virginia*, edited by D. W. Linzey, 483–621. Center for Environmental Studies, Virginia Polytechnic Institute, Blacksburg, VA. Information on the distribution, life history, and status of endangered and threatened mammals of Virginia, with accounts of rare and poorly known species as well.

Handley, C. O., Jr., and C. P. Patton. 1947. *Wild Mammals of Virginia*. Virginia Commission of Game and Inland Fisheries, Richmond, VA. Discussions on several major groups of Virginia mammals, keys, and individual accounts of the wild mammals of Virginia.

Leatherwood, S., D. K. Caldwell, and H. E. Winn. 1976. *Whales, Dolphins, and Porpoises of the Western North Atlantic*. National Marine Fisheries Service, NOAA Technical Report, Circular 396. A technical synopsis of all cetaceans occurring in the North Atlantic Ocean, with many illustrations.

Lee, D. S., and J. B. Funderburg. 1977. Mammals. In *Endangered and Threatened Plants and Animals of North Carolina*, edited by J. E. Cooper, S. S. Robinson, and J. B. Funderburg, North Carolina State Museum of Natural History, Raleigh, NC. A brief introduction to the study of mammals in North Carolina with discussions on the status and distribution of rare, poorly known, threatened, and endangered species.

Lee, D. S., J. B. Funderburg, Jr., and M. K. Clark. 1982. *A Distributional Survey of North Carolina Mammals*. Occasional Papers North Carolina Biological Survey, 1982:10. Raleigh, NC. Brief sections on the history of the study of mammals in North Carolina, mammal habitats, and zoogeography; and brief accounts dealing with the status, distribution, and habitats of North Carolina mammals.

Linzey, A. V., and D. W. Linzey. 1971. *Mammals of the Great Smoky Mountains National Park*. University of Tennessee Press, Knoxville, TN. Accounts of each species of mammal that occurs in the Great Smoky Mountains National Park and a section that relates the park fauna to

that of the surrounding area.

Linzey, D. W., and A. V. Linzey. 1968. Mammals of the Great Smoky Mountains National Park. *Journal of Elisha Mitchell Scientific Society* 84:384–414. A similar but more scientific account than the last entry.

Paradiso, J. L. 1969. *Mammals of Maryland*. North American Fauna, Number 66, Bureau of Sport Fisheries and Wildlife, Washington, DC. Information on Maryland mammal habitats, a series of keys to species identification, and individual accounts of the mammals of Maryland, with distribution maps.

Potter, C. W. 1984. Marine Mammals of Maryland. In *Threatened and Endangered Plants and Animals of Maryland*, edited by A. W. Norden, D. C. Forester, and G. H. Fenwick, 442–53. Maryland Natural Heritage Program, Special Publications 84-1. Information on the distribution and status of marine mammals of Maryland.

Schmidly, D. J. 1981. *Marine Mammals of the Southeastern United States Coast and the Gulf of Mexico*. U.S. Fish and Wildlife Service, Washington, DC. A technical report containing much information on identification, distribution, status, and life history of marine mammals in the region covered.

Smith, E. R., J. B. Funderburg, Jr., and T. L. Quay. 1960. *A Checklist of North Carolina Mammals*. North Carolina Wildlife Resources Commission, Raleigh, NC. A brief listing of the species, distribution, and abundance of the mammals of North Carolina.

General Publications

American Society of Mammalogists. 1969–present. *Mammalian Species.* American Society of Mammalogists, Allen Press, Inc., Lawrence, KS. A series of separate leaflets, each dealing in detail with the biology of a particular mammalian species.

Anderson, S., and J. K. Jones, Jr. 1984. *Orders and Families of Recent Mammals of the World.* John Wiley and Sons, New York, NY. A technical synopsis of mammalian orders and families, with distribution maps.

Barbour, R. W., and W. H. Davis. 1969. *Bats of America.* University Press of Kentucky, Lexington, KY. Information on the identification of bats, with keys and a series of individual accounts of the lives of bats in North America, including range maps.

Bourlière, F. 1964. *The Natural History of Mammals.* 3d edition. Alfred A. Knopf, Inc., New York, NY. A popular but scientific account of the biology of mammals of the world.

Chapman, J. A., and G. A. Feldhamer, eds. 1982. *Wild Mammals of North America: Biology, Management, and Economics.* The Johns Hopkins University Press, Baltimore, MD. Rather detailed accounts of game, fur-bearing species, and other economically important species of mammals of North America, with considerable attention to management.

Gunderson, H. L. 1976. *Mammalogy.* McGraw-Hill Book Company, New York, NY. A college level mammalogy text covering all aspects of the study of mammals but with a strong emphasis on mammalian biology.

Hall, E. R. 1981. *The Mammals of North America.* John Wiley and Sons, New York, NY. Vols. 1-2. Scientific accounts of the mammals of North America providing diagnostic characters and keys to species and subspecies, including range maps.

Honacki, J. H., K. E. Kinman, and J. W. Koeppl. 1982. *Mammal Species of the World: A Taxonomic and Geographic Reference.* Allen Press, Inc., Lawrence, KS. A complete list of all mammal species, their distributions, and other technical information.

Jones, J. K., Jr., D. C. Carter, H. H. Genoways, R. S. Hoffman, and D. W. Rice. 1982. *Revised Checklist of North American Mammals North of Mexico, 1982.* Occasional Papers, The Museum, Texas Tech University, Number 80. A comprehensive list of all mammals occurring north of Mexico with their scientific and common names, including notes on recent changes in nomenclature.

Kurten, B. 1972. *The Age of Mammals.* Columbia University Press, New York, NY. A popular account of the evolutionary history of mammals.

MacDonald, D., ed. 1984. *The Encyclopedia of Mammals.* Facts on File, Inc., New York, NY. Extensive descriptions of many of the mammals of the world, with numerous illustrations.

Nowak, R. M., and J. L. Paradiso. 1983. *Walker's Mammals of the World.* 4th edition. The Johns Hopkins University Press, Baltimore, MD. Vols. 1–2. A series of accounts of the mammal species of the world, including many illustrations.

Vaughan, T. A. 1978. *Mammalogy*. 2d edition. W. B. Saunders Co., Philadelphia, PA. A college level mammalogy text covering all aspects of the study of mammals, but with a strong emphasis on classification.

Field Guides

Burt, W. H., and R. P. Grossenheider. 1976. *A Field Guide to the Mammals*. 3d edition. Houghton Mifflin Co., Boston, MA. Brief descriptions and comments on habitat, habits, and young, with illustrations and range maps.

Glass, B. P. 1973. *A Key to the Skulls of North American Mammals*. Bryan P. Glass, Oklahoma State University, Stillwater. Keys to the identification of mammals from skull characters.

Leatherwood, S., and R. Reeves. 1982. *The Sierra Club Handbook of Whales and Dolphins*. Sierra Club Books, San Francisco, CA. A complete list of all whales and dolphins with illustrations and information on the biology of each species.

Murie, O. J. 1954. *A Field Guide to Animal Tracks*. Houghton Mifflin Co., Boston, MA. Information on natural history, tracks, and other sign left by mammals, including many illustrations.

Smith, R. P. 1982. *Animal Tracks and Signs of North America*. Stackpole Books, Harrisburg, PA. A well illustrated discussion of tracks and sign which will aid the reader in recog-

nizing and interpreting wildlife clues.

Whitaker, J. O., Jr. 1980. *The Audubon Society Field Guide to North American Mammals*. Alfred A. Knopf, Inc., New York, NY. Descriptions of mammals, their sign, habitat, and range; includes many photographs.

Techniques Manuals

Anderson, R. M. 1965. *Methods of Collecting and Preserving Vertebrate Animals*. 4th edition. Bulletin of the Natural Museum of Canada, Number 69, Biol. Ser. 18, Ottawa, Canada. Detailed information on collecting and preserving mammals for museum collections.

Hall, E. R. 1962. *Collecting and Preparing Study Specimens of Vertebrates*. Miscellaneous Publications Number 30, University of Kansas Museum of Natural History, Lawrence, KS. Similar to above.

Current Journals and Magazines

Journal of Mammalogy. Published quarterly by the American Society of Mammalogists, Vertebrate Museum, Shippensburg University, Shippensburg, PA 17257. Results of original research on all aspects of the biology of mammals; society membership is open to all interested persons.

The Journal of Wildlife Management. Published quarterly by the Wildlife Society, Inc., 5410 Grosvenor Lane, Bethesda, MD 20814. Results of original research on the management of animals, including mammals; deals primarily with game species, predators, and endangered or threatened species; society membership is open to all interested persons.

South Carolina Wildlife. Published bimonthly by the South Carolina Wildlife and Marine Resources Department, 1000 Assembly St., Columbia, SC 29201. Often contains informative articles about mammals of the state.

Virginia Wildlife. Published monthly by the Virginia Commission of Game and Inland Fisheries, 4010 W. Broad St., Richmond, VA 23230. Same as above.

Wildlife in North Carolina. Published monthly by the North Carolina Wildlife Resources Commission, 325 N. Salisbury St., Raleigh, NC 27611. Same as above.

Photo Credits

The photographs in this book were taken by Dr. James F. Parnell except for those provided by the following photographers and agencies: Dr. Roger W. Barbour (water shrew, rock vole), Dr. Walter C. Biggs, Jr. (mountain stream, Virginia foothills, chipmunk burrow, pine-cone scales, gray squirrel, black bear, killer whale, white-tailed deer buck, elk, bison), Mr. Peter C. Cram (bottle-nosed dolphin), Dr. Thomas W. French (southeastern shrew), Dr. Donald F. Kapraun (harbor seal), Mr. Dwight R. Kuhn (star-nosed mole), Dr. John R. MacGregor (gray myotis, Keen's myotis, meadow jumping mouse), Mr. Nicky L. Olsen (Brazilian free-tailed bat), Mr. Wayne R. Peterson and Manomet Bird Observatory (sperm whale), Dr. Roger A. Powell (fisher), Mr. William E. Sanderson (nine-banded armadillo, swamp rabbit), Dr. John L. Tveten (Seminole bat, northern yellow bat), Dr. Wm. David Webster (mountain bald, mammal tracks), Ms. Nancy M. Wells (northern flying squirrel), Dr. Howard E. Winn and Cetacean and Turtle Assessment Program (Atlantic spotted dolphin), Cetacean Research Unit—Provincetown Center for Coastal Studies (humpback whale), Great Smoky Mountains National Park (spotted skunk), Texas Parks and Wildlife Department (red wolf), and the U.S. Fish and Wildlife Service (manatee).

Index